T0336892

Construction Hazardous Materials Compliance Guide

Asbestos Detection, Abatement, and Inspection Procedures

R. Dodge Woodson

AMSTERDAM • BOSTON • HEIDELBERG • LONDON
NEW YORK • OXFORD • PARIS • SAN DIEGO
SAN FRANCISCO • SINGAPORE • SYDNEY • TOKYO

Butterworth-Heinemann is an imprint of Elsevier

Butterworth-Heinemann is an imprint of Elsevier
225 Wyman Street, Waltham, MA 02451, USA
The Boulevard, Langford Lane, Kidlington, Oxford, OX5 1GB, UK

Notices
Knowledge and best practice in this field are constantly changing. As new research
and experience broaden our understanding, changes in research methods, professional
practices, or medical treatment may become necessary.

    Practitioners and researchers must always rely on their own experience and knowledge
in evaluating and using any information, methods, compounds, or experiments described
herein. In using such information or methods they should be mindful of their own
safety and the safety of others, including parties for whom they have a professional
responsibility.

    To the fullest extent of the law, neither the Publisher nor the authors, contributors,
or editors, assume any liability for any injury and/or damage to persons or property as a
matter of products liability, negligence or otherwise, or from any use or operation of any
methods, products, instructions, or ideas contained in the material herein.

Library of Congress Cataloging-in-Publication Data
Woodson, R. Dodge (Roger Dodge), 1955 –
    Asbestos detection, abatement, and inspection procedures / R. Dodge Woodson.
        p. cm. — (Construction hazardous materials compliance guide)
    Includes index.
    ISBN 978-0-12-415841-2
    1. Asbestos abatement.   I.   Title.
    TD887.A8W66 2012
    628.5'3—dc23   2011037999

British Library Cataloguing-in-Publication Data
A catalogue record for this book is available from the British Library.

For information on all Butterworth–Heinemann publications
visit our web site at *www.elsevierdirect.com*.

Printed in the United States
12   13   14   15   16   10 9 8 7 6 5 4 3 2 1

This book is dedicated to Giovanni Ivan Sinclair,
my daughter's first baby and my first grandchild.
Welcome to the world, Vanni.

Contents

Acronyms

ACBM: Asbestos-Containing Building Material
ACM: Asbestos-Containing Material
AHERA: Asbestos Hazardous Emergency Response Act
ASHARA: Asbestos School Hazard Abatement Reauthorization Act
DOT: Department of Transportation
EPA: Environmental Protection Agency
HEPA: High-Efficiency Particulate Air
HVAC: Heating, Ventilation, and AirConditioning
LEA: Local Education Agency
MAP: Asbestos Model Accreditation Plan
NESHAP: National Emission Standard for Hazardous Air Pollutants
NIOSH: National Institute of Occupational Safety and Health
O&M: Operations and Maintenance
OSHA: Occupational Safety and Health Administration
PCM: Phase Contrast Microscopy
PLM: Polarized Light Microscopy
SSSD: Small Scale, Short Duration
TEM: Transmission Electron Microscopy
TSI: Thermal System Insulation
VAT: Vinyl Asbestos Tile
VOC: Volatile Organic Compound

Introduction

If you are a contractor who works with older homes and buildings, you may well encounter asbestos building materials. When this happens, many situations can arise. Bidding a job that will require professional asbestos containment or abatement without including the work in your bid can result in a financial disaster.

Should you disturb asbestos in any way, your actions are likely to release asbestos fibers into the surrounding air. There are serious health risks associated with asbestos becoming airborne. Negligence on your part can result in major lawsuits. Asbestos is not a material to be taken lightly, and any older building can contain some form of asbestos.

Whether you are a remodeling contractor, carpenter, plumber, heating mechanic, roofer, siding contractor, or flooring contractor, you can be at risk. Most buildings and homes constructed prior to 1980 are likely sources of asbestos materials.

The laws, regulations, and rules governing working with asbestos are set forth by both federal and state agencies. Any failure to comply with the requirements detailed by governing authorities can result in stiff fines and potential lawsuits. These requirements are extensive and can be complex.

This book provides invaluable insight and guidance in working with asbestos. There is a deep mixture of information at your fingertips. It ranges from Woodson's field experience to rules, regulations, and laws on both state and federal levels.

If you have any chance of encountering asbestos in your work, you need this invaluable guide to keep you informed and safe. Look at the table of contents. Thumb through the pages. You will see quickly that this is a comprehensive guide for all types of contractors. Do not let yourself be taken off guard. Read this book and prepare yourself to deal with asbestos at future job locations. The companion site for this book can be accessed at *booksite.elsevier.com/ 9780124158412.*

▶ ACKNOWLEDGMENTS

I want to thank the following who contributed to my research and facts in the writing of this book:

- OSHA
- EPA
- State of New Hampshire
- State of Maine
- State of Oklahoma
- State of Delaware
- State of Virginia
- CanStockPhoto
- Big Stock Photos

▶ ABOUT THE AUTHOR

R. Dodge Woodson is a career contractor with more than 30 years of experience. He has been a Master Plumber and remodeling contractor since 1979. The Woodson name is synonymous with professional reference books. R. Dodge Woodson has written numerous best-selling books over the years.

What Is Asbestos?

Asbestos is the name given to a number of naturally occurring fibrous minerals with high tensile strength, the ability to be woven, and resistance to heat and most chemicals. Because of these properties, asbestos fibers have been used in a wide range of manufactured goods, including roofing shingles, ceiling and floor tiles, paper and cement products, textiles, coatings, and friction products such as automobile clutch, brake, and transmission parts. The Toxic Substances Control Act defines asbestos as the asbestiform varieties of chrysotile (serpentine), crocidolite (riebeckite), amosite (cummingtonite/grunerite), anthophyllite, tremolite, and actinolite.

The term *asbestos* describes six naturally occurring fibrous minerals found in certain types of rock formations. It is a mineral compound of silicon, oxygen, hydrogen, and various metal cations. Of the six types, the minerals chrysotile, amosite, and crocidolite have been most commonly used in building products. When mined and processed, asbestos is typically separated into very thin fibers. When these fibers are present in the air, they are normally invisible to the naked eye. Asbestos fibers are commonly mixed during processing with a material that binds them together so that they can be used in many different products. Because these fibers are so small and light, they may remain in the air for many hours if they are released from the asbestos-containing material (ACM) in a building.

Asbestos became a popular commercial product to manufacturers and builders in the early 1900s to the 1970s. Asbestos is durable and fire retardant, resists corrosion, and insulates well. It is estimated that 3,000 different types of commercial products contain some amount of asbestos. The use of asbestos ranges from paper products and brake linings to floor tiles and thermal insulation. See Box 1.1 for key points about asbestos. Intact and undisturbed, ACM does not pose a health risk. Asbestos becomes a problem when, due to damage, disturbance, or deterioration over time, the material releases fibers into the air.

Box 1.1. Key Points about Asbestos

This chapter introduces some important terms used in the AHERA Rule. The designated person should be especially familiar with the following:

Asbestos-Containing Material (ACM)—Any material or product that contains more than 1 percent asbestos.

Asbestos-Containing Building Material (ACBM)—Surfacing ACM, thermal system insulation ACM, or miscellaneous ACM that is found in or on interior structural members or other parts of a school building.

Friable ACBM—Material that may be crumbled, pulverized, or reduced to powder by hand pressure when dry. Friable ACBM also includes previously nonfriable material when it becomes damaged to the extent that when dry it may be crumbled, pulverized, or reduced to powder by hand pressure.

Nonfriable ACBM—Material that, when dry, may not be crumbled, pulverized, or reduced to powder by hand pressure.

Surfacing ACM—Interior ACM that has been sprayed on, troweled on, or otherwise applied to surfaces (e.g., structural members, walls, ceilings, etc.) for acoustical, decorative, fireproofing, or other purposes.

Thermal System ACM—Insulation used to control heat transfer or prevent condensation on pipes and pipe fittings, boilers, breeching, tanks, ducts, and other parts of hot and cold water systems; heating, ventilation, and air-conditioning (HVAC) systems; or other mechanical systems.

Miscellaneous ACM—Other, mostly nonfriable, products and materials (found on structural components, structural members or fixtures) such as floor tile, ceiling tile, construction mastic for floor and ceiling materials, sheet flooring, fire doors, asbestos cement pipe and board, wallboard, acoustical wall tile, and vibration damping cloth.

Undamaged nonfriable ACBM should be treated as friable if any action performed would render these materials friable. When previously nonfriable ACBM becomes damaged to the extent that when dry it may be crumbled, pulverized, or reduced to powder by hand pressure, it should be treated as friable.

Source: www.docstoc.com/docs/586090/How-to-Manage-Asbestos-in-School-Buildings-AHERA-Designated-Person-s-Self-study-Guide.

▶ ASBESTOS HEALTH EFFECTS

If inhaled, tiny asbestos fibers can impair normal lung functions and increase the risk of developing lung cancer, mesothelioma, or asbestosis. It could take anywhere from 20 to 30 years after the first exposure for symptoms to occur. Severe health problems from exposure have been experienced by workers who held jobs in industries such as shipbuilding, mining, milling, and fabricating.

Exposure to asbestos increases your risk of developing lung disease. That risk is made worse by smoking. In general, the greater the exposure to asbestos, the greater the chance of developing harmful health effects. Disease symptoms may take several years to develop following exposure. If you are concerned about possible exposure, consult a physician who specializes in lung diseases (pulmonologist).

Exposure to airborne friable asbestos may result in a potential health risk because persons breathing the air may breathe in asbestos fibers. Continued exposure can increase the amount of fibers that remain in the lung. Fibers embedded in lung tissue over time may cause serious lung diseases including asbestosis, lung cancer, or mesothelioma. Smoking increases the risk of developing illness from asbestos exposure.

The following are three of the major health effects associated with asbestos exposure:

- **Asbestosis**—Asbestosis is a serious, progressive, long-term noncancer disease of the lungs. It is caused by inhaling asbestos fibers that irritate lung tissues and cause the tissues to scar. The scarring makes it difficult for oxygen to get into the blood. Symptoms of asbestosis include shortness of breath and a dry, crackling sound in the lungs while inhaling. There is no effective treatment for asbestosis.
- **Lung cancer**—Lung cancer causes the largest number of deaths related to asbestos exposure. People who work in the mining, milling, and manufacturing of asbestos and those who use asbestos and its products are more likely to develop lung cancer than the general population. The most common symptoms of lung cancer are coughing and a change in breathing. Other symptoms include shortness of breath, persistent chest pains, hoarseness, and anemia.
- **Mesothelioma**—Mesothelioma is a rare form of cancer that is found in the thin lining (membrane) of the lung, chest, abdomen, and heart, and almost all cases are linked to exposure to asbestos. This disease may not show up until many years after asbestos exposure. This is why great efforts are being made to prevent schoolchildren from being exposed.

▶ WHERE CAN ASBESTOS BE FOUND?

Asbestos fibers are incredibly strong and have properties that make them resistant to heat. Many products are in use today that contain asbestos. Most of these are materials used in heat and acoustic insulation, fire proofing, and roofing and flooring. In 1989, the Environmental Protection Agency (EPA) identified the following asbestos product categories. Many of these materials may still be in use. Table 1.1 provides examples of such products.

▶ WHAT SHOULD PROPERTY OWNERS DO IF THEY DISCOVER ASBESTOS?

The best thing to do is to leave alone any asbestos-containing material that is in good condition. If unsure whether or not the material contains asbestos, you may consider hiring a professional asbestos inspector to sample and test the material. Before you have your house remodeled, you should find out whether asbestos-containing materials are present.

If asbestos-containing material is becoming damaged (i.e., unraveling, frayed, breaking apart), you should immediately isolate the area (keep pets and children away from the area) and refrain from disturbing the material (either by touching it or walking on it). You should then immediately contact an asbestos professional for consultation.

TABLE 1.1. Materials That May Contain Asbestos

ASBESTOS-CEMENT CORRUGATED SHEET	ASBESTOS-CEMENT FLAT SHEET	ASBESTOS-CEMENT PIPE	ASBESTOS-CEMENT SHINGLES
Roof coatings	Flooring felt	Pipeline wrap	Roofing felt
Asbestos clothing	Non-roof coatings	Vinyl/asbestos floor tile	Automatic transmission components
Clutch facings	Disc brake pads	Drum brake linings	Brake blocks
Commercial and industrial asbestos friction products	Sheet and beater-add gaskets (except specialty industrial)	Commercial, corrugated, and specialty paper	Millboard
Rollboard			

It is best to receive an assessment from one firm and any needed abatement from another firm to avoid any conflict of interest. In such a scenario as described here, asbestos-containing material does not necessarily need to be removed but may rather be repaired by an asbestos professional via encapsulation or enclosure. Removal is often unnecessary.

Laboratory Testing

The National Institute of Standards and Technology (NIST) maintains a listing of accredited asbestos laboratories under the National Voluntary Laboratory Accreditation Program (NVLAP). You may call NIST at 301-975-4016.

How to Identify Materials That Contain Asbestos

You can't tell whether a material contains asbestos simply by looking at it, unless it is labeled. If in doubt, treat the material as if it contains asbestos and have it sampled and analyzed by qualified professionals. A professional should take samples for analysis because the professional knows what to look for and because there may be an increased health risk if fibers are released.

In fact, if done incorrectly, sampling can be more hazardous than leaving the material alone. Taking samples if you are not trained in the proper procedures is not recommended. Material that is in good condition and will not be disturbed should be left alone. Only material that is damaged or will be disturbed should be sampled. Table 1.2 shows a timeline for asbestos regulations.

▶ MANAGING ASBESTOS PROBLEMS

If the asbestos material is in good shape and will not be disturbed, your best option is to do nothing. See Figure 1.1 for an example of shingles that are still in good shape. If the asbestos is a problem, there are two types of corrections. You can either repair or remove the asbestos.

Repair usually involves either sealing or covering asbestos material. This is a fairly common procedure when there is no immediate threat of asbestos exposure.

- Sealing (encapsulation) involves treating the material with a sealant that either binds the asbestos fibers together or coats the material so fibers are not released. Pipe, furnace, and

TABLE 1.2. Timeline of Regulatory and Legislative Activities

1900	Asbestos recognized as a cause of occupational disease (asbestosis) in Charing Cross Hospital, London. A presumptive connection is established.
1918	Insurance companies, including Prudential, refuse to sell insurance to asbestos workers.
1922	U.S. Navy lists asbestos work as hazardous and recommends the use of respirators.
1924	Asbestos is established as a definitive cause of death from lung scarring.
1927	The name "asbestosis" is applied to lung scarring caused by asbestos. Massachusetts awards disability payments to individuals affected by occupational lung disease. Over the next 40 years, other states come to recognize asbestosis as a compensable disease.
1929	Workers begin suing Johns Manville for damages from disability caused by asbestos exposure.
1931	In the United Kingdom, Parliament requires dust-control measures in asbestos textile factories and allows workers to receive compensation for asbestosis. A "safe" level is established as conditions such that no more than one in three workers will get asbestosis after 15 to 19 years work exposure.
1946	The American Conference of Governmental Industrial Hygienists (ACGIH) establishes a maximum acceptable concentration (MAC) in 1946 of 5 million particles per cubic foot (mppcf) for occupational exposure.
1948	The 5 mppcf MAC was changed to a threshold limit value (TLV) of an average concentration over an 8-hour day, referred to as an 8-hour, time-weighted average.
1955	Richard Doll publishes paper linking asbestos to lung cancer.
1960	Chris Wagner publishes paper linking asbestos to mesothelioma.
1964	Johns Manville first places warning labels on some asbestos products. Irving J. Selikoff describes the incidence of asbestos-related disease among insulation workers.
1969	First product-liability lawsuit is brought against asbestos manufacturers. Federal contracts over $10,000 must adhere to a workplace standard of 12 fibers per cubic centimeter (f/cc) of air.
1970	OSHA establishes the first federal guidelines for workplace asbestos exposure.
1971	OSHA regulations take effect. EPA lists asbestos as a hazardous air pollutant.
1972	ACGIH lists asbestos as a human carcinogen. First permanent asbestos regulations instituted by OSHA. Permissible exposure limit (PEL) is 5 f/cc.
1973	First NESHAP rule enacted. Eliminates spray application of fireproofing containing asbestos. Asbestos consumption in United States hits all-time high of over 800,000 tons.
1975	NESHAP revision bans the use of asbestos in many thermal insulation products. EPA defines "friable" asbestos.

TABLE 1.2. *Cont'd*

1976	OSHA reduces PEL to 2 f/cc.
1977	Consumer Products Safety Commission (CPSC) issues rules prohibiting the sale of consumer patching compounds and fireplace emberizing agents containing respirable freeform asbestos.
1978	NESHAP is revised.
1979	EPA begins providing technical assistance to help schools identify and control friable ACM. The primary document is known as the "orange book."
1982	EPA promulgates "Asbestos in Schools" rule.
1983	EPA "orange book" is revised to provide guidance to manage friable asbestos in non-school buildings. The new document is the "blue book."
1984	EPA national survey estimates that there are 733,000 buildings with friable ACM. Asbestos School Hazard Abatement Act passed.
1985	The last comprehensive EPA guidance document for asbestos in buildings is issued: *Guidance for Controlling Asbestos-Containing Materials in Buildings*, also known as the "purple book."
1986	OSHA reduces PEL to 0.2 f/cc, with an "excursion limit" of 1 f/cc for up to 30 minutes. Asbestos Hazard Emergency Response Act (AHERA) is passed.
1986	CPSC issues an enforcement policy under the Federal Hazardous Substances Act (FHSA) concerning labeling of certain asbestos-containing household products.
1987	EPA issues AHERA regulations. EPA promulgates Asbestos Worker Protection Rule, applying OSHA standards to employees of state and local governments.
1989	EPA promulgates Asbestos Ban and Phase-Out Rule.
1990	NESHAP is revised. Asbestos School Hazard Abatement Reauthorization Act passed. EPA holds policy dialog with stakeholders regarding asbestos in public and commercial buildings. The "green book," a guidance document on operations and maintenance programs for the management of in-place ACM, is issued.
1991	Much of the Ban and Phase-Out Rule is vacated by the U.S. Circuit Court of Appeals. The portion prohibiting new uses for asbestos remains intact. Health Effects Institute compiles *Asbestos Research on Asbestos in Public and Commercial Buildings*, a review and synthesis of the literature.
1991	EU bans amphibole asbestos. Chrysotile is banned for some applications. Chief Justice Rehnquist of the U.S. Supreme Court appoints an ad hoc committee regarding the thousands of court-filed asbestos illness claims.
1992	EPA attempts to work with auto industry to voluntarily phase out asbestos in brakes. Threatened anti-trust action by asbestos industry ends this effort.

(Continued)

TABLE 1.2. Timeline of Regulatory and Legislative Activities—*cont'd*

1994	OSHA reduces PEL to 0.1 f/cc. Under this OSHA standard, thermal system insulation (TSI) and surfacing materials installed before 1981, and floor tile installed through 1981, are presumed to be asbestos-containing materials unless demonstrated otherwise through sampling.
1999	EU extends ban on chrysotile to nearly all applications. Member states must enact bans by 2005.
2000	Asbestos Worker Protection Rule is revised.
2002	"Ban Asbestos in America Act" is introduced by U.S. Senator Patty Murray (D-WA).

Source: http://gelmans.com/ReadingRoom/tabid/65/ctl/ArticleView/mid/372/articleId/362/Timeline-of-Asbestos-Regulatory-and-Legislative-Activities.aspx.

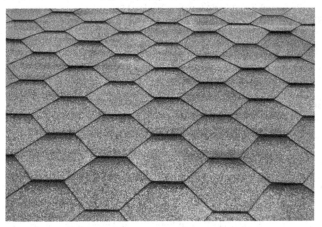

Figure 1.1 Asbestos shingles in good shape.

boiler insulation can sometimes be repaired this way. This procedure should be done only by a professional trained to handle asbestos safely.

• Covering (enclosure) involves placing something over or around the material that contains asbestos to prevent release of fibers. Exposed insulated piping may be covered with a protective wrap or jacket.

With any type of repair, the asbestos remains in place. Repair is usually cheaper than removal, but it may make later removal of asbestos, if necessary, more difficult and costly. Repairs can either be major or minor.

Asbestos Dos and Don'ts for Homeowners

As a contractor who works with asbestos, you may need to educate potential customers as part of your job. This can be a good way to build a potential customer base. For example, you might offer a free seminar where you discuss the risks of asbestos and the proper procedures for property owners to adopt in an effort to avoid exposure to asbestos. The people attending your seminar may become customers.

At the very least, you will have brought the subject to light and educated the public in what to look out for and what to do if they suspect they may have a problem with asbestos. The following are some key topics that you should provide to property owners:

- Do keep activities to a minimum in any areas having damaged material that may contain asbestos.
- Do take every precaution to avoid damaging asbestos material.
- Do have removal and major repair done by people who are trained and qualified in handling asbestos. It is highly recommended that sampling and minor repair also be done by asbestos professionals.
- Don't dust, sweep, or vacuum debris that may contain asbestos.
- Don't saw, sand, scrape, or drill holes in asbestos materials.
- Don't use abrasive pads or brushes on power strippers to strip wax from asbestos flooring. Never use a power stripper on a dry floor.
- Don't sand or try to level asbestos flooring or its backing. When asbestos flooring needs replacing, install new floor covering over it, if possible.
- Don't track material that could contain asbestos through the house. If you cannot avoid walking through the area, have it cleaned with a wet mop. If the material is from a damaged area, or if a large area must be cleaned, call an asbestos professional.
- Major repairs must be done only by a professional trained in methods for safely handling asbestos.
- Minor repairs should also be done by professionals because there is always a risk of exposure to fibers when asbestos is disturbed.
- Doing minor repairs yourself is not recommended because improper handling of asbestos materials can create a hazard where none existed.

Removal is usually the most expensive method and, unless required by state or local regulations, should be the last option considered in most situations. The reason is that removal poses the greatest risk of fiber release. However, removal may be required when remodeling or making major changes to your home that will disturb asbestos material. Also, removal may be called for if asbestos material is damaged extensively and cannot be otherwise repaired. Removal is complex and must be done only by a contractor with special training. Improper removal may actually increase the health risks to you and your family.

▶ ASBESTOS PROFESSIONALS

Asbestos professionals are trained in handling asbestos material. The type of professional will depend on the type of product and what needs to be done to correct the problem. You may hire a general asbestos contractor or, in some cases, a professional trained to handle specific products containing asbestos. Roofers are one example of such specialists, as shown in Figure 1.2.

Asbestos professionals can conduct home inspections, take samples of suspected material, assess its condition, and advise about what corrections are needed and who is qualified to make the corrections. Once again, material in good condition need not be sampled unless it is likely to be disturbed. Professional correction or abatement contractors repair or remove asbestos materials.

Figure 1.2 An asbestos worker abating a roof that contained asbestos.

Some firms offer combinations of testing, assessment, and correction. A professional hired to assess the need for corrective action should not be connected with an asbestos-correction firm. It is better to use two different firms so there is no conflict of interest. Services vary from one area to another around the country.

The federal government has training courses around the country for asbestos professionals. Some state and local governments also have or require training or certification courses. Ask asbestos professionals to document their completion of federal or state-approved training. Each person performing work in your home should provide proof of training and licensing in asbestos work, such as completion of EPA-approved training. State and local health departments or EPA regional offices may have listings of licensed professionals in your area.

If you have a problem that requires the services of asbestos professionals, check their credentials carefully. Hire professionals who are trained, experienced, reputable, and accredited—especially if accreditation is required by state or local laws. Before hiring a professional, ask for references from previous clients. Find out whether they were satisfied. Ask whether the professional has handled similar situations. Is the contractor insured? When you act as the general contractor, it is up to you to make certain that you are getting true professionals to handle your asbestos needs. Get cost estimates from several professionals because the charges for these services can vary.

Though private homes are usually not covered by the asbestos regulations that apply to schools and public buildings, professionals should still use procedures described during federal or state-approved training. You should be alert to the chance of misleading claims by asbestos consultants and contractors.

There have been reports of firms incorrectly claiming that asbestos materials in homes must be replaced. In other cases, firms have encouraged unnecessary removals or performed them improperly. Unnecessary removals are a waste of money. Improper removals may actually increase the health risks. To guard against this, know what services are available and what procedures and precautions are needed to do the job properly.

In addition to general asbestos contractors, you may select a roofing, flooring, or plumbing contractor trained to handle asbestos when it is necessary to remove and replace roofing, flooring, siding, or asbestos-cement pipe that is part of a water system. Normally, roofing and flooring contractors are exempt from state and local licensing requirements because they do not perform any other asbestos-correction work. Before you authorize anyone or any trade to work with asbestos, you should be sure that the workers are qualified to perform the necessary work.

The following list points out suggestions for what you should look for in hiring professionals who work with asbestos and air quality:

- Make sure that the inspection will include a complete visual examination and the careful collection and lab analysis of samples. If asbestos is present, the inspector should provide a written evaluation describing its location and extent of damage, and give recommendations for correction or prevention.
- Make sure an inspecting firm makes frequent site visits if it is hired to assure that a contractor follows proper procedures and requirements. The inspector may recommend and perform checks after the correction to assure the area has been properly cleaned.
- Check with your local air pollution control board, the local agency responsible for worker safety, and the Better Business Bureau. Ask if the firm has had any safety violations. Find out if there are legal actions filed against it.
- Insist that the contractor use the proper equipment to do the job. Workers must wear approved respirators, gloves, and other protective clothing (see Figure 1.3).
- Before work begins, get a written contract specifying the work plan; cleanup; and the applicable federal, state, and local regulations that the contractor must follow (such as notification requirements and asbestos disposal procedures). Contact your state and local health departments, EPA regional office, and the OSHA regional office to find out what the regulations are. Be sure the contractor follows local asbestos removal and disposal laws. At the end of the job, get written assurance from the contractor that all procedures have been followed.
- Assure that the contractor avoids spreading or tracking asbestos dust into other areas of your home. The contractor should seal the work area from the rest of the house using plastic sheeting and duct tape, and also turn off the heating and

Figure 1.3 This asbestos worker is wearing protective clothing.

air-conditioning system. For some repairs, such as pipe insulation removal, plastic glove bags may be adequate. They must be sealed with tape and properly disposed of when the job is complete.

- Make sure the work site is clearly marked as a hazard area. Do not allow building occupants or their pets into the work area until work is completed.
- Insist that the contractor apply a wetting agent to the asbestos material with a hand sprayer that creates a fine mist before removal. Wet fibers do not float in the air as easily as dry fibers and will be easier to clean up.
- Make sure the contractor does not break removed material into small pieces. This could release asbestos fibers into the air. Pipe insulation was usually installed in preformed blocks and should be removed in complete pieces.
- Upon completion of the project, assure that the contractor cleans the area well with wet mops, wet rags, sponges, or high-efficiency particulate air (HEPA) vacuum cleaners. A regular

vacuum cleaner must never be used. Wetting helps reduce the chance of spreading asbestos fibers in the air. All asbestos materials and disposable equipment and clothing used in the job must be placed in sealed, leak-proof, and labeled plastic bags. The work site should be visually free of dust and debris. Air monitoring (to make sure there is no increase of asbestos fibers in the air) may be necessary to assure that the contractor's job is done properly. This should be done by someone not connected with the contractor.

• Do not dust, sweep, or vacuum debris that may contain asbestos. These steps will disturb tiny asbestos fibers and may release them into the air. Remove dust by wet mopping or with a special HEPA vacuum cleaner used by trained asbestos contractors.

▶ REGULATIONS GOVERNING ASBESTOS

There are a multitude of regulations governing asbestos. The following list identifies most of the sources of asbestos regulations:

TSCA: The Toxic Substances Control Act first authorized the EPA to regulate asbestos in schools and public and commercial buildings under Title II of the law, also known as the Asbestos Hazard Emergency Response Act (AHERA).

AHERA: The Asbestos Hazard Emergency Response Act requires local education agencies (LEAs) to inspect their schools for ACBM and prepare management plans to reduce the asbestos hazard. The act also established a program for the training and accreditation of individuals performing certain types of asbestos work.

ASHARA: The Asbestos School Hazard Abatement and Reauthorization Act reauthorized AHERA and made some minor changes to the act. It also reauthorized the Asbestos School Hazard Abatement Act.

ASHAA: The Asbestos School Hazard Abatement Act of 1984 provided loans and grants to help financially needy public and private schools correct serious asbestos hazards. This program was funded from 1985 until 1993. There have been no funds appropriated since that date.

CAA-Asbestos NESHAP: Pursuant to the Clean Air Act (CAA) of 1970, the EPA established the asbestos National Emission Standard for Hazardous Air Pollutants (NESHAP). It is intended to minimize the release of asbestos fibers during activities involving the handling of asbestos. It specifies work practices to be followed during renovation, demolition, or other abatement activities when friable asbestos is involved.

Here are some key facts to remember about asbestos:

- Although asbestos is hazardous, human risk of asbestos disease depends on exposure.
- Removal is often not the best course of action to reduce asbestos exposure. In fact, an improper removal can create a dangerous situation where none previously existed.
- The EPA requires removal only to prevent significant public exposure to asbestos, such as during building renovation or demolition.
- The EPA recommends in-place management whenever asbestos is discovered. Instead of removal, implementation of a management plan will usually control fiber release when materials are not significantly damaged and are not likely to be disturbed.
- If you have any questions concerning asbestos or would like to report illegal or improper asbestos activity, contact: TSCA Hotline: 202-554-1404; Asbestos Hotline: 800-368-5888; The Regional Asbestos Coordinator: 214-665-7575; or The NESHAP Asbestos Coordinator: 214-665-7575.

Now that you know what asbestos is, it's time to get down to working with it. There is plenty to learn, so let's get busy.

Overview of Codes, Regulations, and Standards

2

This chapter will give you a good overview of the codes, regulations, and standards set forth for those who will work with asbestos products. As we move through the process in the following chapters, I will give you more specific details and instructions. Tables 2.1 and 2.2 show examples of the types of products to look for.

Keep in mind that codes, laws, and regulations change from time to time; see an example of codes, regulations, and standards in Figure 2.1. This information is based on my current knowledge and research. However, it is wise to use the information here as a foundation of knowledge and check current local requirements prior to commencing any work. Now, let's get down to business.

TABLE 2.1. Asbestos-Containing Products Allowed in the United States

• asbestos cement corrugated sheet	• clutch facings
• asbestos cement flat sheet	• friction materials
• asbestos clothing	• disk brake pads
• pipeline wrap	• drum brake lining
• roofing felt	• brake blocks
• vinyl asbestos floor tile	• gaskets
• asbestos cement shingle	• non-roofing coatings
• millboard	• roof coatings
• asbestos cement pipe	• automatic transmission components

TABLE 2.2. Asbestos-Containing Products Not Subject to 1989 Ban and Phase-Out

• acetylene cylinders	• arc chutes
• asbestos diaphragms	• battery separators
• high-grade electrical paper	• missile liners
• packings (valves, seals, and other uses)	• reinforced plastic
• sealant tape	• specialty industrial gaskets
• textile products	

PART 1 – GENERAL

RELATED DOCUMENTS

General provisions of Contract and other Division-1 Specification Sections apply to this section.

SUMMARY

This section sets forth governmental regulations and industry standards which are included and incorporated herein by reference and made a part of the specification. This section also sets forth those notices and permits which are known to the Owner and which either must be applied for and received, or which must be given to governmental agencies before start of work.

- Requirements include adherence to work practices and procedures set forth in applicable codes, regulations, and standards.
- Requirements include obtaining permits, licenses, inspections, releases, and similar documentation, as well as payments, statements, and similar requirements associated with codes, regulations, and standards.

CODES AND REGULATIONS

General Applicability of Codes and Regulations, and Standards: Except to the extent that more explicit or more stringent requirements are written directly into the contract documents, all applicable codes, regulations, and standards have the same force and effect (and are made a part of the contract documents by reference) as if copied directly into the contract documents, or as if published copies are bound herewith.

Contractor Responsibility: The Contractor shall assume full responsibility and liability for the compliance with all applicable federal, state, and local regulations pertaining to work practices, hauling, disposal, and protection of workers, visitors to the site, and persons occupying areas adjacent to the site. The Contractor is responsible for providing medical examinations and maintaining medical records of personnel as required by the applicable federal, state, and local regulations. The Contractor shall hold the government and contracting officer harmless for failure to comply with any applicable work, hauling, disposal, safety, health, or other regulation on the part of himself, his employees, or his subcontractors.

Federal Requirements: which govern asbestos abatement work or hauling and disposal of asbestos waste materials using the most current edition which include but are not limited to the following:

OSHA: U.S. Department of Labor, Occupational Safety and Health Administration (OSHA) using most current edition and incorporating future additions, including but not limited to:

> Occupational Exposure to Asbestos, Tremolite,
> Anthophyllite, and Actinolite; Final Rules
> Title 29, Part 1910, Section 1001, and Part 1926, Section 1101 of the Code of Federal Regulations

Figure 2.1 Section 01092, Codes, Regulations, and Standards—Asbestos Abatement

Respiratory Protection
Title 29, Part 1910, Section 134 of the Code of Federal Regulations

Construction Industry
Title 29, Part 1926 of the Code of Federal Regulations

Access to Employee Exposure and Medical Records
Title 29, Part 1910, Section 2 of the Code of Federal Regulations

Hazard Communication
Title 29, Part 1910, Section 1200 of the Code of Federal
Regulations

Specifications for Accident Prevention Signs and Tags
Title 29, Part 1910, Section 145 of the Code of Federal Regulations

DOT: U.S. Department of Transportation, including but not limited to:

Hazardous Substances
Title 29, Parts 171 and 172 of the Code of Federal Regulations

EPA: U.S. Environmental Protection Agency (EPA), including but not
limited to:

Asbestos Abatement Projects; Worker Protection Rule
Title 40, Part 763, Subpart G of the Code of Federal Regulations

Asbestos Hazard Emergency Response Act (AHERA) Regulation
Asbestos-Containing Materials in Schools Final Rule & Notice
Title 40, Part 763, Subpart E of the Code of Federal Regulations

Training Requirements of (AHERA) Regulation:

Asbestos Containing Materials in Schools Final Rule & Notice
Title 40, Part 763, Subpart E, Appendix C of the Code of Federal
Regulations

National Emission Standard for Hazardous Air Pollutants
(NESHAPS)
National Emission Standard for Asbestos
Title 40, Part 61, Subparts A and M of the Code of Federal
Regulations

State Requirements: which govern asbestos abatement work or hauling and
disposal of asbestos waste materials include but are not limited to the local
codes, regulations, and standards in place at the time work is to be done.

STANDARDS
General Applicability of Standards: Except to the extent that more explicit
or more stringent requirements are written directly into the Contract
Documents, all applicable standards have the same force and effect (and
are made a part of the Contract Documents by reference) as if copied
directly into the Contract Documents, or as if published copies are bound
herewith.

Figure 2.1 *Cont'd*

Contractor Responsibility: The Contractor shall assume full responsibility and liability for the compliance with all standards pertaining to work practices, hauling, disposal, and protection of workers, visitors to the site, and persons occupying areas adjacent to the site. The Contractor shall hold the government and contracting officer harmless for failure to comply with any applicable standard on the part of himself, his employees, or his subcontractors.

Standards: which apply to asbestos abatement work or hauling and disposal of asbestos waste materials include but are not limited to the following:

American National Standards Institute (ANSI)
1430 Broadway, New York, NY 10018
(212) 354-3300

Fundamentals Governing the Design and Operation of Local Exhaust Systems, Publication Z9.2-79

Practices for Respiratory Protection, Publication Z88.2-80

American Society for Testing and Materials (ASTM)
1916 Race Street, Philadelphia, PA 19103
(215) 299-5400

Safety and Health Requirements Relating to Occupational Exposure to Asbestos, E849-82`

Specification for Encapsulants for Friable Asbestos-Containing Building Materials

EPA GUIDANCE DOCUMENTS

Discuss asbestos abatement work or hauling and disposal of asbestos waste materials listed below for the Contractor's information only. These documents do not describe the work and are not a part of the work of this contract. EPA maintains an information number, (800) 334-8571 and publications can be ordered from (800) 424-9065 (202-554-1404 in Washington, DC):

- Asbestos-Containing Materials in School Buildings—A Guidance Document, Parts 1 & 2 (Orange Books)
- EPA C00090 (out of print)
- Guidance for Controlling Asbestos-Containing Materials in Buildings (Purple Book), EPA 560/5-85-024
- Friable Asbestos-Containing Materials in Schools: Identification and Notification Rule (40 CFR Part 763)
- Evaluation of the EPA Asbestos-in-Schools Identification and Notification Rule, EPA 560/5-84-005
- Asbestos in Buildings: National Survey of Asbestos-Containing Friable Materials, EPA 560/5-84-006
- Asbestos in Buildings: Guidance for Service and Maintenance Personnel, EPA 560/5-85-018
- Asbestos Waste Management Guidance, EPA 530-SW-85-007

Figure 2.1 *Cont'd*

- Asbestos Fact Book, EPA Office of Public Affairs
- Asbestos in Buildings, Simplified Sampling Scheme for Friable Surfacing Materials
- Commercial Laboratories with Polarized Light Microscopy Capabilities for Bulk Asbestos Identification
- A Guide to Respiratory Protection for the Asbestos Abatement Industry, EPA-560-OPTS-86-001

NOTICES

U.S. ENVIRONMENTAL PROTECTION AGENCY

Send Written Notification: as required by USEPA National Emission Standards for Hazardous Air Pollutants (NESHAPS) Asbestos Regulations (40 CFR 61, Subpart M) to the regional Asbestos NESHAPS contact at least 10 days prior to beginning any work on asbestos-containing materials. Send notification to the following address:

REGION 6
Asbestos NESHAPS Contact
Air & Waste Management Division
USEPA
1201 Elm Street
Dallas, TX 75270
(214) 767-9835

Notification: Include the following information in the notification sent to the NESHAPS contact:

- Name and address of owner or operator.
- Description of the facility being demolished or renovated, including the size, age, and prior use of the facility.
- Estimate of the approximate amount of friable asbestos material present in the facility in terms of linear feet of pipe, and surface area on other facility components. For facilities in which the amount of friable asbestos materials less than 80 linear meters (260 linear feet) on pipes and less than 15 square meters (160 square feet) on other facility components explain techniques of estimation.
- Location of the facility being demolished or renovated.
- Scheduled starting and completion dates of demolition or renovation.
- Nature of planned demolition or renovation and method(s) to be used.
- Procedures to be used to comply with the requirements of USEPA NESHAPS Asbestos Regulations (40 CFR 61, Subpart M).
- Name and location of the waste disposal site where the friable asbestos waste material will be deposited.
- For facilities being demolished under an order of a state or local governmental agency, issued because the facility is structurally unsound and in danger of imminent collapse, the name, title, and authority of the state or local governmental representative who has ordered the demolition.

Figure 2.1 *Cont'd*

STATE AND LOCAL AGENCIES

Send written notification as required by state and local regulations prior to beginning any work on asbestos-containing materials.

LICENSES

Maintain current licenses as required by applicable state or local jurisdictions for the removal, transporting, disposal, or other regulated activity relative to the work of this contract.

POSTING AND FILING OF REGULATIONS

Post all notices required by applicable federal, state, and local regulations. Maintain two (2) copies of applicable federal, state, and local regulations and standard. Maintain one copy of each at job site. Keep on file in Contractor's office one copy of each.

Figure 2.1 *Cont'd*

▶ SUMMARY

In summary, you will encounter plenty of rules and regulations when you work with asbestos. Keeping up with them is your responsibility. See Box 2.1 for an operations and management plan. Failure to do so can be expensive in fines and personal injury. Now that you have a general overview of the business, we can get down to the details.

Box 2.1. Key Points about the Operations and Management Program

An O&M program must be implemented whenever any **friable** ACBM is present or assumed to be present in a school building or whenever any nonfriable ACBM or assumed nonfriable ACBM is about to become friable as a result of activities performed in the school building.

Unless the building has been cleaned using similar methods in the previous 6 months, all areas of a building where friable ACBM, friable suspected ACBM assumed to be ACBM, or significantly damaged TSI ACBM is present must be **cleaned** using the methods described at §763.91(c) of the AHERA Rule at least once after the completion of the AHERA inspection and before the initiation of any response action, other than O&M activities or repair.

Specialized work practices and procedures must be followed for any O&M activities disturbing **friable** ACBM.

Box 2.1. *Cont'd*

When a fiber release episode occurs, the work practices that must be followed depend on whether the episode is minor or major in nature. A **minor fiber release episode** consists of the falling or dislodging of 3 square or linear feet or less of friable ACBM. A **major fiber release episode** consists of the falling or dislodging of more than 3 square or linear feet of friable ACBM.

Once ACBM is identified or assumed to be present, the LEA should start a **notification and warning program** to alert affected parties to a potential hazard in the building and to provide basic information on how to avoid the hazard.

The LEA is required to attach a **warning label** immediately adjacent to any friable and nonfriable ACBM and suspected ACBM that is assumed to be ACBM that is located in routine maintenance areas.

Where employees work in areas where fiber levels exceed permissible exposure limits or are required to wear pressure respirators, the LEA must establish **medical surveillance and respiratory protection programs.**

A designated person can minimize accidental disturbances of ACBM during maintenance and renovation activities by establishing a **permit system** that calls for all work orders and requests to be processed through the designated person.

The specific work practices that must be followed when routine maintenance activities are being conducted depend on the likelihood that the activities will disturb the ACBM and cause fibers to be released.

Identifying Asbestos

3

All tested materials that contain 1 percent asbestos fibers or more, using the polarized light microscopy (PLM) method, are considered hazardous. The disturbance or dislocation of such asbestos-containing materials (ACM) may cause asbestos fibers to be released into the environment, thereby creating a potential health hazard to workers and building occupants. The general project goal is to identify cost-effective means of dealing with ACM that comply with all applicable regulations and rules and minimize health and environmental risks during the asbestos abatement, removal, or disturbance activities.

The designer needs to identify those materials that may be disturbed during construction and thus may be potential sources of friable asbestos. For example, on a heating job, the pipe insulation should be tested, as well as flooring or walls that may be penetrated by heating pipes. Similarly, on an electrical job, areas of conduit penetration should be tested.

Materials installed prior to 1980 are classified as presumed asbestos-containing materials (PACM). This presumption can be rebutted by testing using the PLM method. Insulation materials, resilient floor tiles, roofing and siding shingles, roofing felts, mastics, joint compounds, caulking, glazing, gaskets, wall boards, and transite panels are the usual ACM used in building construction.

▶ DESIGN

All asbestos abatement work should take place in accordance with the provisions outlined in the current local, state, and federal regulations. In particular, work must adhere to Massachusetts Department of Labor and Workforce Development (DLWD) and Department of Environmental Protection (DEP) regulations regarding asbestos removal and disposal.

After the locations of the asbestos-containing materials have been determined, the design goal is the selection of the appropriate, cost-effective abatement methods. In general, the options are removal, encapsulation, or management in place. Figure 3.1 shows the use of plastic to allow the removal of an asbestos ceiling.

Figure 3.1   Room enclosed with plastic to allow removal of asbestos ceiling.

▶ ASBESTOS-CONTAINING MATERIALS

ACM can be placed into the following categories:

1. **Category I:** *Friable Asbestos-Containing Material (Friable ACM)* is defined as any material that contains more than 1 percent asbestos, which, when dry, may be crumbled, pulverized, or reduced to powder by hand pressure. It also includes nonfriable ACM when such material becomes damaged to the extent that, when dry, it may be crumbled, pulverized, or reduced to powder by hand pressure.

   Class I Asbestos work generally involves the removal/disturbance of thermal system insulation (TSI) and surfacing ACM or PACM. This removal procedure requires a full containment and a three-stage decontamination unit under negative pressure. Clearance air sampling at the end of the asbestos removal is mandatory.

2. **Category II:** *Nonfriable Asbestos-Containing Material* is defined as any material excluding Category I friable ACM that contains more than 1 percent asbestos.

   Class II Asbestos work generally involves the removal/disturbance of nonfriable ACM that is not thermal system insulation or surfacing material. According to the definition, this includes, but is not limited to, the removal of asbestos-containing wallboard, floor tiles, sheeting, gaskets, joint compounds, roofing felts, roofing and siding shingles, and construction mastics. The work area should be properly isolated to prevent release of asbestos fibers into the adjacent spaces or into the environment. The contractor should be required to erect mini containments and use wetting agents during the removal/disturbance of ACM. Visual inspection at the end of the asbestos removal process is mandatory.

3. **Class III** Asbestos work generally involves the removal of small amounts of Category II materials using a glove-bag method or other alternative methods for small-scale removals and/or disturbances.

▶ SAMPLE SITUATIONS

Let's consider scenarios where asbestos is a problem. Assume that what you are about to read is a description of real cases.

A Basement Conversion

You have been asked to convert a full basement into finished living space. A site visit as part of your estimating process reveals old basement piping that is covered with deteriorated asbestos pipe insulation that needs to be removed, discarded, and replaced with new insulation. The white covering and banding around the asbestos have come loose. Asbestos is clearly visible. This just will not do. Some form of action is needed to secure the asbestos.

In this case, this scope typically falls under the Class I removal procedure and requires a full containment abatement method, with a three-stage decontamination unit under negative pressure. It is not going to be a cheap process, but it is needed. Any affected area should be posted with warning signs, as is shown in Figure 3.2.

Sometimes there are simple solutions to dealing with asbestos, but this is not one of those cases. If the insulation had been present and intact, you could have escaped costs by repairing or encapsulating the material. Because the material is deteriorated and exposed, a major cleanup is required.

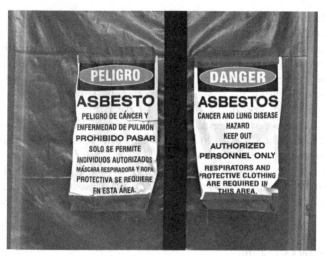

Figure 3.2 Enclosed asbestos work area with required signage.

Kitchen Remodeling

You are doing an estimate to remodel a kitchen. This is what we in the trade call a "gut-and-go job." In other words, you will be stripping the room down to the subfloor, wall studs, and ceiling joists and then recreating it with new products. There is a problem. You have found that the existing vinyl flooring contains asbestos. The good news is you recognized the conditions before bidding the job. But now you have to bring in an asbestos professional to access the situation.

Vinyl asbestos tile (VAT) can be partially abated to accommodate new floor penetrations (e.g., for heating and electrical system upgrade projects) as Class III maintenance scope. However, this job requires much more than mere penetrations. You will be doing a complete removal, and this increases the requirements that you must follow. This job calls for a full-scale abatement.

▶ ALTERNATIVE APPROACHES

Alternative approaches exist for dealing with various ACMs; the designer's task is to identify the method that best balances the budget, environmental risk, and longevity. The time required for residents to be out of their unit should be taken into consideration when determining the appropriate method of ACM removal. Every effort should be made to minimize the relocation time required.

Requirements for procedures during abatement are defined by the applicable regulations; however, it is important to note that asbestos removal under full containment is not the only procedure allowed by regulations.

Contract documents must clearly identify the type and provide quantities of asbestos-containing materials to be abated. They should also identify existing conditions that will affect the work of the abatement contractor such as location of electric panels and water lines that will be used for temporary services, proposed locations of HEPA exhaust systems and decontamination facilities, and so on. Coordination shall exist between the abatement under this section and the work of other trades. It is important that the contract documents be written to allow the contractor to decide how to complete the work using the most cost-effective, compliant work practice.

▶ ASBESTOS WASTE

Asbestos-containing waste shall be containerized, transported, and disposed in compliance with all local and state regulations. Full-time abatement monitoring is not required for asbestos abatement projects. Normal construction administration services, with the parallel services of the asbestos abatement consultant, including conducting the initial submittal reviews (e.g., medical records, licenses), final visual inspections, and air clearance tests (whichever is required by the class of abatement), are usually adequate project oversight.

At the conclusion of the abatement process, the asbestos consultant should submit a detailed report to the Landholding Agency (LHA); the report needs to include a summary of abatement operations, results of air sampling, and documentation relative to the proper disposal of asbestos waste.

Health Facts about Asbestos

4

Asbestos is the name given to a group of minerals that occur naturally in the environment as bundles of fibers that can be separated into thin, durable threads. These fibers are resistant to heat, fire, and chemicals and do not conduct electricity. For these reasons, asbestos has been used widely in many industries.

Chemically, asbestos minerals are silicate compounds, meaning they contain atoms of silicon and oxygen in their molecular structure.

Asbestos minerals are divided into two major groups: serpentine asbestos and amphibole asbestos. Serpentine asbestos includes the mineral chrysotile, which has long, curly fibers that can be woven. Chrysotile asbestos is the form that has been used most widely in commercial applications.

Amphibole asbestos includes the minerals actinolite, tremolite, anthophyllite, crocidolite, and amosite. Amphibole asbestos has straight, needle-like fibers that are more brittle than those of serpentine asbestos and are more limited in their ability to be fabricated.

Since the late 1800s, asbestos has been mined and used commercially in North America. Its use increased greatly during World War II. Since then, asbestos has been used in many industries. For example, the building and construction industries have used it for strengthening cement and plastics as well as for insulation, roofing, fireproofing, and sound absorption.

The shipbuilding industry has used asbestos to insulate boilers, steam pipes, and hot water pipes. The automotive industry uses asbestos in vehicle brake shoes and clutch pads. Asbestos has also been used in ceiling and floor tiles; paints, coatings, and adhesives; and plastics. In addition, asbestos has been found in vermiculite-containing garden products and some talc-containing crayons.

In the late 1970s, the U.S. Consumer Product Safety Commission (CPSC) banned the use of asbestos in wallboard patching compounds and gas fireplaces because the asbestos fibers in these products could be released into the environment during use. In addition, manufacturers of electric hairdryers voluntarily stopped using asbestos in their products in 1979. In 1989, the U.S. Environmental Protection Agency (EPA) banned all new uses of asbestos; however, uses developed before 1989 are still allowed. The EPA also established regulations that require school systems to inspect buildings for the presence of damaged asbestos and to eliminate or reduce asbestos exposure to occupants by removing the asbestos or encasing it.

In June 2000, the CPSC concluded that the risk of children's exposure to asbestos fibers in crayons was extremely low. However, U.S. manufacturers of these crayons agreed to eliminate talc from their products.

In August 2000, the EPA conducted a series of tests to evaluate the risk for consumers of adverse health effects associated with exposure to asbestos-contaminated vermiculite. The EPA concluded that exposure to asbestos from some vermiculite products poses only a minimal health risk. The EPA recommended that consumers reduce the low risk associated with the occasional use of vermiculite during gardening activities by limiting the amount of dust produced. Specifically, the EPA suggested that consumers use vermiculite outdoors or in a well-ventilated area; keep vermiculite damp while using it; avoid bringing dust from vermiculite into the home on clothing; and use premixed potting soil, which is less likely to generate dust.

The regulations described before and other actions, coupled with widespread public concern about the health hazards of asbestos, have resulted in a significant annual decline in the U.S. use of asbestos. Domestic consumption of asbestos amounted to about 803,000 metric tons in 1973, but it had dropped to about 2,400 metric tons by 2005.

▶ WHAT ARE THE HEALTH HAZARDS OF EXPOSURE TO ASBESTOS?

People may be exposed to asbestos in their workplace, their communities, or their homes. If products containing asbestos are disturbed, tiny asbestos fibers are released into the air. When asbestos fibers are breathed in, they may get trapped in the lungs and remain there for a long time. Over time, these fibers can accumulate and cause scarring and inflammation, which can affect breathing and lead to serious health problems. Some of the health risks are shown in Box 4.1.

> ### Box 4.1. Key Points about Asbestos Health Risks
>
> Asbestos-related diseases are dose-response related (the greater the exposure to airborne fibers, the greater the risk of developing an illness) and have a latency period (typically 15 to 30 years).
>
> Exposure to asbestos may result in **asbestosis** (a disease characterized by lung scarring, which reduces the lungs' ability to function), **lung cancer**, **mesothelioma** (always-fatal cancer arising in the chest or abdominal cavity), and **other diseases.**
>
> Risks associated with low-level, nonoccupational exposure (e.g., a building occupant who is not actually disturbing the asbestos) are not well established. The National Institute for Occupational Safety and Health (NIOSH) has determined, however, that there is no established safe level of exposure.
>
> Asbestos that has been identified will pose little risk if it is well maintained under an operations and maintenance program. EPA *only* requires asbestos removal to prevent significant public exposure to airborne asbestos fibers during building demolition or renovation activities.

Asbestos has been classified as a known human carcinogen (a substance that causes cancer) by the U.S. Department of Health and Human Services (HHS), the EPA, and the International Agency for Research on Cancer. Studies have shown that exposure to asbestos may increase the risk of lung cancer and mesothelioma (a relatively rare cancer of the thin membranes that line the chest and abdomen).

Even though rare, mesothelioma is the most common form of cancer associated with asbestos exposure. In addition to lung cancer and mesothelioma, some studies have suggested an association between asbestos exposure and gastrointestinal colorectal cancers, as well as an elevated risk for cancers of the throat, kidney, esophagus, and gallbladder. However, the evidence is inconclusive.

Asbestos exposure may also increase the risk of asbestosis (an inflammatory condition affecting the lungs that can cause shortness of breath, coughing, and permanent lung damage) and other nonmalignant lung and pleural disorders, including pleural plaques (changes in the membranes surrounding the lung), pleural thickening, and benign pleural effusions (abnormal collections of fluid between the thin layers of tissue lining the lungs and the wall of the chest cavity). Although pleural plaques are not precursors to lung cancer, evidence suggests that people with pleural disease caused by exposure to asbestos may be at increased risk for lung cancer.

▶ WHO IS AT RISK FOR AN ASBESTOS-RELATED DISEASE?

Everyone is exposed to asbestos at some time during his or her life. Low levels of asbestos are present in the air, water, and soil. However, most people do not become ill from their exposure. People who become ill from asbestos are usually those who are exposed to it on a regular basis, most often in a job where they work directly with the material or through substantial environmental contact.

Since the early 1940s, millions of American workers have been exposed to asbestos. Health hazards from asbestos fibers have been recognized in workers exposed in the shipbuilding trades, asbestos mining and milling, manufacturing of asbestos textiles and other asbestos products, insulation work in the construction and building trades, and a variety of other trades. Demolition workers, drywall removers, asbestos removal workers, firefighters, and automobile workers also may be exposed to asbestos fibers.

Studies evaluating the cancer risk experienced by automobile mechanics exposed to asbestos through brake repair are limited, but the overall evidence suggests there is no safe level of asbestos exposure. As a result of government regulations and improved work practices, today's workers (those without previous exposure) are likely to face smaller risks than those exposed in the past.

Individuals involved in the rescue, recovery, and cleanup at the site of the September 11, 2001, attacks on the World Trade Center (WTC) in New York City are another group at risk of developing an asbestos-related disease. Because asbestos was used in the construction of the North Tower of the WTC, when the building was attacked, hundreds of tons of asbestos were released into the atmosphere.

Those at greatest risk include firefighters, police officers, paramedics, construction workers, and volunteers who worked in the rubble at Ground Zero. Others at risk include residents in close proximity to the WTC towers and those who attended schools nearby. These individuals will need to be followed to determine the long-term health consequences of their exposure.

One study found that nearly 70 percent of rescue and recovery workers suffered new or worsened respiratory symptoms while performing work at the site. The study describes the results of the Laborers' Health & Safety Fund of North America WTC Worker and Volunteer Medical Screening Program, which was established to identify and characterize possible WTC-related health effects in responders. It was found that about 28 percent of those tested had abnormal lung function tests, and 61 percent of those without previous health problems developed respiratory symptoms. However, it is important to note that these symptoms may be related to exposure to debris components other than asbestos.

Although it is clear that the health risks from asbestos exposure increase with heavier exposure and longer exposure time, investigators have found asbestos-related diseases in individuals with only brief exposures. Generally, those who develop asbestos-related diseases show no signs of illness for a long time after their first exposure. It can take from 10 to 40 years or more for symptoms of an asbestos-related condition to appear.

There is some evidence that family members of workers exposed to asbestos regularly face an increased risk of developing mesothelioma. This risk is thought to result from exposure to asbestos fibers brought into the home on workers' shoes, clothing, skin, and hair. To decrease these exposures, federal law regulates workplace practices to limit the possibility of asbestos being brought home in this way. Some employees may be required to shower and change clothes before they leave work, store street clothes in a separate area of the workplace, or wash their work clothes at home separately from other clothes.

Cases of mesothelioma have also been seen in individuals without occupational asbestos exposure who live close to asbestos mines.

▶ WHICH FACTORS AFFECT THE RISK OF DEVELOPING AN ASBESTOS-RELATED DISEASE?

Several factors can help to determine how asbestos exposure affects an individual, including:

- Dose (how much asbestos an individual was exposed to).
- Duration (how long an individual was exposed).
- Size, shape, and chemical makeup of the asbestos fibers.

- Source of the exposure.
- Individual risk factors, such as smoking and preexisting lung disease.

Although all forms of asbestos are considered hazardous, different types of asbestos fibers may be associated with different health risks. For example, the results of several studies suggest that amphibole forms of asbestos may be more harmful than chrysotile, particularly for mesothelioma risk, because they tend to stay in the lungs for a longer period of time.

► HOW DOES SMOKING AFFECT RISK?

Many studies have shown that the combination of smoking and asbestos exposure is particularly hazardous. Smokers who are also exposed to asbestos have a risk of developing lung cancer that is greater than the individual risks from asbestos and smoking added together. There is evidence that quitting smoking will reduce the risk of lung cancer among asbestos-exposed workers. Smoking combined with asbestos exposure does not appear to increase the risk of mesothelioma. However, people who were exposed to asbestos on the job at any time during their life or who suspect they may have been exposed should not smoke.

► HOW ARE ASBESTOS-RELATED DISEASES DETECTED?

Individuals who have been exposed (or suspect they have been exposed) to asbestos fibers on the job, through the environment, or at home via a family contact should inform their doctor about their exposure history and whether they experience any symptoms. The symptoms of asbestos-related diseases may not become apparent for many decades after the exposure. It is particularly important to check with a doctor if any of the following symptoms develop:

- Shortness of breath, wheezing, or hoarseness.
- A persistent cough that gets worse over time.
- Blood in the sputum (fluid) coughed up from the lungs.
- Pain or tightening in the chest.
- Difficulty swallowing.
- Swelling of the neck or face.
- Loss of appetite.
- Weight loss.
- Fatigue or anemia.

A thorough physical examination, including a chest X-ray and lung function tests, may be recommended. The chest X-ray is currently the most common tool used to detect asbestos-related diseases. However, it is important to note that chest X-rays cannot detect asbestos fibers in the lungs, but they can help identify any early signs of lung disease resulting from asbestos exposure.

Studies have shown that computed tomography (CT)—a series of detailed pictures, created by a computer linked to an X-ray machine, of areas inside the body taken from different angles—may be more effective than conventional chest X-rays at detecting asbestos-related lung abnormalities in individuals who have been exposed to asbestos.

A lung biopsy, which detects microscopic asbestos fibers in pieces of lung tissue removed by surgery, is the most reliable test to confirm the presence of asbestos-related abnormalities. A bronchoscopy is a less invasive test than a biopsy and detects asbestos fibers in material that is rinsed out of the lungs. It is important to note that these tests cannot determine how much asbestos an individual may have been exposed to or whether disease will develop. Asbestos fibers can also be detected in urine, mucus, or feces, but these tests are not reliable for determining how much asbestos may be in an individual's lungs.

▶ HOW CAN WORKERS PROTECT THEMSELVES FROM ASBESTOS EXPOSURE?

The Occupational Safety and Health Administration (OSHA) is a component of the U.S. Department of Labor (DOL) and is the federal agency responsible for health and safety regulations in maritime, construction, manufacturing, and service workplaces. OSHA established regulations dealing with asbestos exposure on the job—specifically in construction work, shipyards, and general industry—that employers are required to follow.

In addition, the Mine Safety and Health Administration (MSHA), another component of the DOL, enforces regulations related to mine safety. Workers should use all protective equipment provided by their employers and follow recommended workplace practices and safety procedures. For example, NIOSH-approved respirators that fit properly should be worn by workers when required.

Workers who are concerned about asbestos exposure in the workplace should discuss the situation with other employees, their employee health and safety representative, and their employers. If necessary, OSHA can provide more information or make an inspection. Regional offices of OSHA are listed in the "United States Government" section of a telephone directory's blue pages (under "Department of Labor"). Regional offices can also be found at *www.osha.gov/html/RAmap.html*.

More information about asbestos is available on the OSHA Asbestos web page, which has links to information about asbestos in the workplace, including what OSHA standards apply, the hazards of asbestos, evaluating asbestos exposure, and controls used to protect workers. This page is available at *www.osha.gov/SLTC/asbestos/index.html*.

OSHA's national office can be contacted at Office of Public Affairs, Occupational Safety and Health Administration, U.S. Department of Labor, Room N-3649, 200 Constitution Avenue, NW, Washington, DC 20210. Use any one of these telephone numbers: 202-693-1999, 1-800-321-6742 (1-800-321-OSHA), or for deaf or hard-of-hearing callers, 1-877-889-5627. The website is *www.osha.gov/workers.html* (workers' page).

Mine workers can contact MSHA at Office of Public Affairs, Mine Safety and Health Administration, U.S. Department of Labor, 21st Floor, 1100 Wilson Boulevard, Arlington, VA 22209. Use either of these telephone numbers: 202-693-9400 or 1-800-746-1553. The website is *www.msha.gov* or *www.msha.gov/codeaphone/codeaphonenew.htm* (National Hazard Reporting Page).

The National Institute for Occupational Safety and Health, which is part of the Centers for Disease Control and Prevention (CDC), is another federal agency that is concerned with asbestos exposure in the workplace. NIOSH conducts asbestos-related research, evaluates work sites for possible health hazards, and makes exposure control recommendations. In addition, NIOSH distributes publications on the health effects of asbestos exposure and can suggest additional sources of information.

NIOSH can be contacted at Education and Information Division, Information Resources Branch, National Institute for Occupational Safety and Health, 4676 Columbia Parkway, Cincinnati, OH 45226; telephone: 1-800-232-7636 (1-800-CDC-INFO); email: cdcinfo@cdc.gov. The website is *www.cdc.gov/niosh*.

▶ WHICH PROGRAMS ARE AVAILABLE TO HELP INDIVIDUALS WITH ASBESTOS-RELATED DISEASES?

Some people with asbestos-related illness may be eligible for Medicare coverage. Information about benefits is available from Medicare's regional offices, located in 10 major cities across the United States and serving specific geographic areas. The regional offices serve as the agency's initial point of contact for beneficiaries, health care providers, state and local governments, and the general public. Contact information for each regional office can be found at *www.cms.hhs.gov/regionaloffices*. General information about Medicare is available by calling toll-free 1-800-633-4227 (1-800-MEDICARE) or by visiting *www.medicare.gov*.

People with occupational asbestos-related diseases also may qualify for financial help, including medical payments, under state workers' compensation laws. Because eligibility requirements vary from state to state, workers employed by private companies or by state and local government agencies should contact their state workers' compensation board. Contact information for state workers' compensation officials may be found in the blue pages of a local telephone directory or at *www.dol.gov/owcp/owcpkeyp.htm*.

If exposure occurred during employment with a federal agency, medical expenses and other compensation may be covered by the Federal Employees' Compensation Program, which is administered by the DOL, Employment Standards Administration's Office of Workers' Compensation Programs.

This program provides workers' compensation benefits to federal (civilian) employees for employment-related injuries and diseases. Benefits include wage replacement, payment for medical care, and, where necessary, medical and vocational rehabilitations assistance in returning to work. Benefits may also be provided to dependents if the injury or disease causes the employee's death. The program has 12 nationwide district offices.

In addition, the Longshore and Harbor Workers' Compensation Program provides benefits to longshoremen, harbor workers, other maritime workers, and other classes of private industry workers who are injured during the course of employment or suffer from diseases caused or worsened by conditions of employment.

Information about eligibility and how to file a claim for benefits under either of these programs can be obtained from Office of Workers' Compensation Programs, Employment Standards Administration, U.S. Department of Labor, Frances Perkins Building, 200 Constitution Avenue, NW, Washington, DC 20210; telephone: 1-866-692-7487 (1-866-OWCPIVR) or 202-693-0040; email: OWCP-Public@dol.gov. The website is *www. dol.gov/owcp*.

Eligible veterans may receive health care at a Department of Veterans Affairs (VA) Medical Center for an asbestos-related disease. Veterans can receive treatment for service-connected and nonservice-connected medical conditions. Information about eligibility and benefits is available from the VA Health Benefits Service Center at 1-877-222-8387 (1-877-222-VETS) or on the VA website at *www1.va.gov/health/index.asp*.

▶ IS THERE FEDERAL LEGISLATION TO HELP VICTIMS OF ASBESTOS-RELATED DISEASES?

No federal legislation has been enacted to compensate victims of asbestos-related diseases or to protect people from asbestos exposure. However, a bill called the Fairness in Asbestos Injury Resolution Act, or FAIR Act, has been introduced in Congress several times. It would create a national trust fund to compensate victims suffering from asbestos-related diseases.

The proposed trust fund would be administered by the DOL, outside the courts, through a claims process in which all individuals with certain medical symptoms and evidence of asbestos-related disease would be compensated. Funding for the trust would come from insurance companies and companies that mined, manufactured, and sold asbestos or asbestos products. Under the bill, individuals affected by asbestos exposure would no longer be able to pursue awards for damages in any federal or state court.

▶ WHICH ORGANIZATIONS OFFER INFORMATION RELATED TO ASBESTOS EXPOSURE?

The organizations listed next can provide more information about asbestos exposure.

The Agency for Toxic Substances and Disease Registry (ATSDR) is the principal federal agency responsible for evaluating the human health effects of exposure to hazardous substances. This agency works in close collaboration with local, state, and other federal agencies; with tribal governments; and with communities and local health care providers to help prevent or reduce harmful human health effects from exposure to hazardous substances.

The ATSDR provides information about asbestos and where to find occupational and environmental health clinics. The agency can be contacted at Agency for Toxic Substances and Disease Registry, 4770 Buford Highway, NE, Atlanta, GA 30341; telephone: 1-800-232-4636 (1-800-CDC-INFO) or 1-888-232-6348 (TTY); email: cdcinfo@cdc.gov. The website is *www.atsdr.cdc.gov*.

The U.S. EPA regulates the general public's exposure to asbestos in buildings, drinking water, and the environment. The EPA offers a Toxic Substances Control Act (TSCA) Hotline and an Asbestos Ombudsman. The TSCA Hotline provides technical assistance and information about asbestos programs implemented under the TSCA, which include the Asbestos School Hazard Abatement Act and the Asbestos Hazard Emergency Response Act. The asbestos ombudsman focuses on asbestos in schools and handles questions and complaints. Both the TSCA Hotline and the asbestos ombudsman can provide publications on a number of topics, particularly on controlling asbestos exposure in schools and other buildings. The Ombudsman operates a toll-free hotline for small businesses, trade associations, and others seeking free, confidential help.

The Environmental Protection Agency's website includes a list of EPA regional and state asbestos contacts at *www.epa.gov/asbestos/pubs/regioncontact.html*. In addition, the EPA's Asbestos and Vermiculite home page provides information about asbestos and its health effects and links to asbestos resources, including suggestions for homeowners who suspect asbestos in their homes, and laws and regulations applicable to asbestos. This page can be found at *www.epa.gov/asbestos*.

Questions may be directed to U.S. Environmental Protection Agency, EPA West Building, National Program Chemicals Division, Mail Code 7404T, 1200 Pennsylvania Avenue, NW, Washington, DC 20460; TSCA hotline: 202-554-1404 or 202-554-0551 (TTY); email: tsca-hotline@epa.gov. For the asbestos ombudsman, call 1-800-368-5888; the website is *www.epa.gov/asbestos*.

Another EPA resource that may be of interest is the brochure titled *Current Best Practices for Preventing Asbestos Exposure among Brake and Clutch Repair Workers*. Released in April 2007, this brochure includes work practices for both automotive professionals and home mechanics that may be used to avoid asbestos exposure. It also summarizes existing OSHA regulatory requirements for professional auto mechanics. The brochure can be found at *www.epa.gov/asbestos/pubs/brakesbrochure.html*.

The CPSC is responsible for protecting the public from unreasonable risks of serious injury or death from more than 15,000 types of consumer products, including asbestos, under the agency's jurisdiction. The CPSC maintains a toll-free 24-hour hotline where callers can obtain product safety and other agency information and report unsafe products.

In addition, CPSC publications provide guidelines for repairing and removing asbestos, and general information about asbestos in the home. The CPSC can be contacted at Office of Information and Public Affairs, U.S. Consumer Product Safety Commission, 4330 East West Highway, Bethesda, MD 20814; telephone: 1-800-638-2772 or 1-800-638-8270 (TTY). The website is *www.cpsc.gov*. Individuals can also contact their local or state health department with questions or concerns about asbestos.

The information here gives a solid overview of health issues connected to asbestos. It is not a comprehensive guide to medical issues, but it does provide a solid foundation on which to build your knowledge. Additional information can be obtained from the sources provided in this chapter.

Quick Tips for Contractors Working with Asbestos

5

The full rules, regulations, and laws pertaining to working with asbestos are complex. They are typically understood and used by professionals who are certified and licensed to work with asbestos. This rarely includes the common general contractor, remodeler, plumber, and so forth. You will find plenty of depth on rules from the Occupational Safety and Health Administration (OSHA), the Environmental Protection Agency (EPA), and an example for the state of Maine in Chapters 8, 9, and 10, respectively. However, you might not be interested in knowing about microns and other detailed data, so this chapter shows the working hardhat the basics of working with asbestos.

The information I give you here is based on the state of New Hampshire. Basically, you should not mess with asbestos if you are not trained and certified to do so. However, if you're a remodeling contractor or general contractor, chances are high that you will encounter building materials that contain asbestos in quantities that will require professional treatment.

Do not use the material here to do your own asbestos work. Also, understand that this chapter provides the fast, down-and-dirty go-to list for general construction and remodeling workers.

▶ ASBESTOS IN HOMES AND BUILDINGS

New Hampshire considers asbestos waste as a solid waste requiring special handling, which coincides with federal policy. The impact on homeowners and land owners in New Hampshire today is that property that contains any type of asbestos is deemed by public sentiment to have certain liabilities. An estimated 80 percent of all buildings constructed before 1978 contain asbestos materials.

Figure 5.1 shows an example of asbestos roofing. It can be found in the following materials, which are listed merely as examples of asbestos-containing materials (ACMs). This list should not be construed as being all-inclusive.

- Siding
- Ceiling tiles

Figure 5.1 A close-up of asbestos roofing.

- Floor tiles
- Roofing materials
- Spray-applied insulation
- Mastic and adhesives
- Pipe insulation
- Linoleum
- Plaster

▶ DEFINITIONS OF ASBESTOS-CONTAINING MATERIAL

Friable ACM means any material that contains more than 1 percent asbestos and can be crumbled, pulverized, or reduced to powder by hand pressure. Nonfriable ACM means any material that contains more than 1 percent asbestos and cannot be pulverized under hand pressure. Nonfriable ACM is divided into two categories. Category I includes packings, gaskets, resilient floor covering, and asphalt roofing products. Category II is any nonfriable ACM not included in Category I. See Chapter 3 for more detail about the categories.

Regulated asbestos-containing material (RACM) includes the following:

- Friable asbestos material.
- Category I nonfriable ACM that has become friable.
- Category I nonfriable ACM that will be, or has been, subjected to sanding, grinding, cutting, or abrading.

- Category II nonfriable ACM that has a high probability of becoming, or has become, crumbled, pulverized, or reduced to powder by the forces expected to act on the material in the course of demolition or renovation operations.

▶ SPECIFIC GUIDANCE FOR CONTRACTORS

With one exception, RACM can be removed only by a licensed asbestos abatement contractor. The one exception to this is an individual homeowner with no tenants doing work on his own private single-family home. Removal of nonregulated asbestos materials can be legally performed by homeowners, regular contractors, and licensed asbestos abatement contractors as long as each does not violate the National Emissions Standards for Hazardous Air Pollutants (NESHAP), which in laymen's terms means no visible emissions, and the work complies with the OSHA's regulations delineated in 29 CFR 1926.1101.

When conducting remediation projects, contractors must be especially careful not to allow employees to

1. Throw asbestos waste material onto hard surfaces where the material will fracture and release fibers.
2. Use techniques such as hammering, sanding, grinding or drilling, which can render nonfriable asbestos materials friable.
3. Remove floor tile using methods that require mechanical chipping or grinding if the tile or the adhesive or mastic contains asbestos fibers.
4. Follow work practices or techniques that will create an environment where the binder asbestos material is disturbed, fractured, or powdered, because such action will create a situation in which fibers will be unlocked and released as airborne particles.

All asbestos-removal projects, whether they involve RACM or nonregulated asbestos, require the project to be supervised by a competent person as defined in 29 CFR 1926.1101.

Handling Tips

The New Hampshire Department of Environmental Services (hereafter called the Department), using experience gained during a number of remediation projects, has developed the following suggestions that we hope will be of use in project implementation:

1. Devise a means of handling and removing nonfriable asbestos material to control the release of fibers and keep breakage to a minimum.
2. Mist the materials being removed with a water spray to prevent fibers from becoming airborne. To enhance the effectiveness of the spray, the water should be treated with a chemical wetting agent. Figure 5.2 shows an example of this.
3. Use plastic or polyethylene (some type of synthetic membrane) to collect errant pieces of material dislodged during removal; for example, use plastic sheeting around the perimeter of a house during siding removal.
4. Package the asbestos waste material as soon as possible and do not leave it unattended or open to the public.
5. Wear a disposable Tyvek suit, gloves, and a half-mask respirator with high-efficiency particulate air (HEPA) filters while removing, packaging, and disposing of the asbestos waste material. It should be noted that it is required to seek a medical opinion before wearing a respirator, which places an additional strain on the heart and lungs.
6. Generate a Waste Shipment Record (WSR). This information/form is required prior to disposal at a facility permitted for that purpose.

Figure 5.2 Asbestos workers are wetting asbestos-containing material.

7. All transporters/haulers must be certain that all loads are secured prior to transport. There is a 1-pound reportable spill quantity for friable asbestos, which requires not only cleaning up the spill but also notifying both the National Response Center (NRC) at 1-800-424-8802 and the New Hampshire Department of Safety (DOS) at 1-800-346-4009.

Insulation

Heating contractors should take note that the guidelines for removing RACM from heating systems are included in the New Hampshire Revised Statutes Annotated (RSA) Chapter 141-E. The removal or repair of RACM from workplaces, schools, public facilities, and dwellings (multiple nonprivate) must be performed by a licensed asbestos abatement contractor. Because this requirement also applies to certain types of ceiling tiles that contain asbestos, general contractors should also take notice. Special exceptions during emergencies allow for owner participation in the removal process. Plastic wrapping of the asbestos allows for isolation of the fibers, basically encapsulating them, while repairs are made or a replacement unit installed.

Replacing and Removing Pipe

Special consideration should be given to situations involving the replacement/removal of transite asbestos pipe and other asbestos piping systems in the ground, above ground, and under bridges. If the pipe is not removed, but a replacement line is run nearby or parallel to the asbestos pipe, the location of the original pipe should be marked on an as-built drawing of the site. Copies should be sent to the Department, Waste Management Division, and to the local code enforcement officer. Future land use and other construction projects may be affected.

Pipe removal projects require the supervision of a competent person and must be performed in accordance with the New Hampshire Solid Waste Rules. Pipe to be removed should not be uprooted by blasting. If sections must be cut in the field to facilitate removal, wetting techniques must be employed.

Dry cutting with a portable radial saw should be avoided. Pipe to be discarded must be packaged in leak-tight containers or sealed in double layers of 6-millimeter-thick plastic and disposed of at a permitted landfill. Care should be expended to reduce the amount of breakage during removal because fracturing causes fiber release. Any pipe crushed and left in place must be located on an as-built drawing.

Removing Floor Coverings

Based on the repromulgated NESHAP (November 20, 1990), the removal of Category I nonfriable resilient asbestos containing floor coverings is regulated as follows:

1. The removal of Category I nonfriable resilient asbestos containing floor coverings does not have to be performed by a licensed asbestos abatement contractor as long as the following conditions are met:
   a. The removal is performed in a manner that does not invoke the NESHAP (see #2), and complies with OSHA regulations delineated in 29 CFR 1926.1101.
   b. The removed material is immediately and properly packaged in leak-tight labeled containers for disposal.
   c. The workers performing removal tasks wear, at a minimum, respiratory protection, protective clothing, and gloves.
   d. The removed asbestos material and contaminated protective equipment is disposed of at a facility that is permitted to accept that type of asbestos waste.
2. Approved removal methodologies that will not invoke the NESHAP, by allowing the tiles to be removed with a minimum of damage to the tiles, are as follows:
   a. Use of heat from heat guns or electric heat machines
   b. Use of infrared machines
   c. Flooding with water or amended water
   d. Use of dry ice or liquid nitrogen
3. Removal techniques that result in severely damaging or fragmenting the asbestos material are unacceptable due to the potential for uncontrolled fiber release. Examples of this would include, but not be limited to, the following activities:
   a. Mechanical chipping
   b. Sanding
   c. Grinding
   d. Abrading
4. Finally, it should be emphasized that some of the mastics used to adhere the resilient flooring coverings to the subfloor

contain asbestos. Therefore, mastic removal techniques must be carefully performed in accordance with approved techniques as just outlined.

Underground Storage Tank Removal

Registered, unregistered, leaking, abandoned, and previously unknown underground storage tanks often had insulated supply and return lines connecting them to a facility, pumphouse, or product line. Approval from the Department must be obtained prior to starting tank removal projects. If the tank is to be made vapor-free and closed in place, an as-built drawing should be generated showing the presence of the asbestos on the underground supply lines. A copy of this drawing must be sent to the Department.

There are other projects that require special attention. See the following list for examples of such situations:

- Asbestos-wrapped cable in underground trenches
- Asphaltic asbestos roofing
- Solid asbestos roofing
- Asphaltic asbestos siding
- Solid asbestos siding
- Asbestos/lead paint

▶ SITE SAFETY AND CONTINGENCY PLANS SUMMARY

Site Safety and Contingency Plans (SSACPs) must be submitted to the Solid Waste Compliance Section (SWCS) for review prior to remediation of asbestos-contaminated properties or outdoor asbestos renovation/demolition/remediation projects. The SSACP, after being reviewed by the SWCS of the Waste Management Division, should be required reading for all personnel connected with the abatement activities and should be posted at the job site.

Emergency telephone numbers should be listed for the job and also posted at the site. These should include, but not be limited to, the local fire department, ambulance service, property owner, contractor's home number, and the nearest hospital. Agencies that are often responsible for monitoring and controlling asbestos are noted in Table 5.1.

TABLE 5.1. Agencies Responsible for Monitoring and Controlling Asbestos

AGENCY	AREA OF RESPONSIBILITY
U.S. Environmental Protection Agency	Products (TSCA), emissions, buildings
Occupational Safety and Health Administration	Workplace products
U.S. Department of Transportation	Shipping
U.S. Food and Drug Administration	Asbestos in foods, drugs, and cosmetics
Mine Safety and Health Administration	Asbestos (during mining)
U.S. Consumer Product Safety Commission	Asbestos in consumer products
U.S. Department of Commerce	Import/export, with EPA
Bureau of Customs and Border Protection	Importation of products, with CPSC

Minimum Requirements

1. A clear description of the project to be undertaken. The description should include location, amount, and type or types of asbestos material to be dealt with and any site-specific conditions.
2. Who will be doing the work? Is the contractor licensed or unlicensed, or is the property owner of record doing the work? Who is the OSHA-competent person designated for the job? A licensed abatement contractor must have up-to-date credentials. A competent person must have current training certificates as well as an up-to-date supervisor's certificate.
3. How will the abatement workers be protected? Is air monitoring required? If so, are both area and personal sampling plans needed? Is an onsite decontamination facility required, or can multiple asbestos-resistant worksuits be used? Has there been a negative initial exposure assessment performed?
4. What provisions are made for crowd control? Can sidewalk superintendents be kept at a safe distance from the work site? Is site security necessary?
5. How is the asbestos waste going to be handled and where is it going? If asbestos materials are to be capped onsite, then as-built drawings will be required. If the wastes are going to a landfill, is it a facility permitted to accept asbestos waste?

If the asbestos waste is to disposed outside the state of New Hampshire, then documentation indicating that it is going to a landfill permitted for that purpose will have to be sent to the Department. Copies of the current operating license and the portion of the operating permit that indicates that the particular type or types of asbestos to be disposed are a waste permitted at that specific facility must be sent to the Department.

6. What will be required for a successful conclusion to the job? Specific information should be furnished detailing the conditions of job approval. This might include, but not be limited to, onsite inspection by multiple parties, documentation-test results, as-built drawings, record drawings, landfill receipts, and WSRs.

7. SSACP are logged in and, after review, abatement activities are allowed to commence. At the conclusion of the abatement activities, submission of air monitoring data, landfill records, and WSRs must be sent to the Department to complete the current project file.

The information here is obviously based on one state. Each state can have its own version of rules, regulations, and laws. In addition to state requirements, contractors must follow guidelines from OSHA and the EPA when working with asbestos. Most important, do not disturb asbestos materials unless you are properly trained and certified to do the work at hand.

Uncertified Contractors' Guide to Inspections and Management

6

Not all contractors are trained or certified for asbestos containment or removal. In fact, most of them are not. This isn't just about general contractors. Remodeling contractors, siding installers, flooring installers, plumbers, heating mechanics, and others all come into play. Any of these groups could come across asbestos in their daily work. The wrong action could produce dangerous results. (See Box 6.1.)

I will assume that you are a building or remodeling contractor who is the general contractor and responsible for entire jobs. How are you going to protect yourself from the health risks of asbestos and possible lawsuits and fines if you make a mistake? Having suitable insurance is a good start, but that is not enough. You need to arm yourself with at least a general working knowledge of asbestos and the role it may have in your professional life.

One way of doing this is to read up on the subject of working with and around asbestos (refer to Table 10.1). A great way is to get the training and certification needed to become licensed to work with asbestos. But, many contractors do not want to perform asbestos containment or abatement. At a minimum, you need to know how to evaluate a job site before you place a bid on a job or do any work. Ignorance can become very expensive in this venue.

Very little is needed in the way of tools when inspecting for asbestos. A flashlight may be the only tool you need in your toolbox. The most important tool is between your ears. The amount of experience and knowledge that you possess is what is most important for a cursory inspection. For example, a building constructed prior to 1980 has a high probability of containing asbestos materials.

Hiring a professional asbestos inspector is a good idea if you are remodeling a building that was built prior to 1984. Asbestos has been used in a wide variety of building materials. If you want to be safe, hire a certified, licensed insurance inspector to go over any suspicious building that you will be working with (see Box 6.2).

Box 6.1. Key Points about the AHERA Inspection

An Asbestos Hazard Emergency Response Act (AHERA) inspection must be conducted by an **accredited inspector.**

The inspector must identify all **homogeneous areas** of material that are suspected to contain asbestos. Homogeneous areas contain asbestos that is uniform (alike) in color and texture.

All material suspected to be asbestos-containing building material (ACBM) must be assumed to be ACBM unless the homogeneous area is **sampled,** and the analysis of the samples shows them to be nonasbestos. An adequate number of samples must be taken or the area will be considered to be ACBM regardless of the results of the analyses.

Once the inspector has identified all ACBM in a building, he or she must perform a **physical assessment** of all thermal system insulation (TSI) and friable ACBM. This involves categorizing the material into one of seven physical assessment classifications.

The results of an AHERA inspection and the assessment must be documented in an **inspection report.** This report will be used by the management planner to make written recommendations on appropriate response actions.

Box 6.2. Key Points about Related Regulations

An asbestos management program is subject not only to AHERA and the AHERA Rule, but also may be subject to **NESHAP, OSHA,** and **DOT** regulations, and the **EPA Worker Protection Rule.**

Relevant provisions of NESHAP establish **work practices for asbestos air emission control** when a facility is being demolished or renovated, and for the disposal of **asbestos-containing waste material.**

The OSHA established **minimum standards for the protection of workers involved in asbestos-related work or employees exposed to asbestos-contaminated workplaces.** These standards include required work practices, engineering controls, permissible exposure limits, written programs for respiratory protection and medical surveillance, methods for compliance, hazard communication, housekeeping, competent person training and responsibilities, and required recordkeeping. OSHA excludes federal, state, or local government employees from its worker protection rules (including public school employees).

The EPA Worker Protection Rule **extends the protection afforded by OSHA** to all employees in asbestos abatement who may have been excluded from protection by OSHA.

Relevant provisions of DOT regulations establish **labeling, packaging, and shipping standards** for the transporting of asbestos-containing materials.

Asbestos pipe insulation is pretty easy to identify. The siding on a building may show obvious signs of being made with asbestos to an experienced contractor. Roofing materials can also be easy to identify, as can some floor coverings. However, there are plenty of situations in which the composition of the material is not obvious to an untrained eye. If you hire a suitable inspector, you hedge the bets in your favor.

Most products made today do not contain asbestos. The few products made that still contain asbestos that could be inhaled are required to be labeled as such. However, until the 1970s, many types of building products and insulation materials used in homes contained asbestos. Common products that might have contained asbestos in the past, and conditions that may release fibers, include the following:

- Steam pipes, boilers, and furnace ducts, insulated with an asbestos blanket or asbestos paper tape. These materials may release asbestos fibers if damaged, repaired, or removed improperly.
- Resilient floor tiles (vinyl asbestos, asphalt, and rubber), the backing on vinyl sheet flooring, and adhesives used for installing floor tile. Sanding tiles can release fibers. So may scraping or sanding the backing of sheet flooring during removal.
- Cement sheet, millboard, and paper used as insulation around furnaces and woodburning stoves. Repairing or removing appliances may release asbestos fibers. So may cutting, tearing, sanding, drilling, or sawing insulation.
- Door gaskets in furnaces, wood stoves, and coal stoves. Worn seals can release asbestos fibers during use.
- Soundproofing or decorative material sprayed on walls and ceilings. Loose, crumbly, or water-damaged material may release fibers; so will sanding, drilling, or scraping the material.
- Patching and joint compounds for walls and ceilings, and textured paints. Sanding, scraping, or drilling these surfaces may release asbestos.
- Asbestos cement roofing, shingles, and siding. These products are not likely to release asbestos fibers unless sawed, drilled, or cut.
- Artificial ashes and embers that are sold for use in gas-fired fireplaces.

▶ HOW TO IDENTIFY MATERIALS THAT CONTAIN ASBESTOS

You can't really tell if a material contains asbestos by looking at it. Some materials are labeled as containing asbestos. Many are not. If there is any doubt, you must assume that asbestos is a potential risk. A professional should take samples for analysis, since he or she knows what to look for and because there may be an increased health risk if fibers are released. In fact, if done incorrectly, sampling can be more hazardous than leaving the material alone. Taking samples yourself is not recommended.

If you choose to take the samples yourself, take care not to release asbestos fibers into the air or onto yourself. Material that is in good condition and will not be disturbed (by remodeling, for example) should be left alone. Only material that is damaged or will be disturbed should be sampled. Anyone who samples asbestos-containing materials should have as much information as possible on the handling of asbestos before sampling and, at a minimum, should observe the following procedures:

- Make sure no one else is in the room when sampling is done.
- Wear disposable gloves or wash hands after sampling.
- Shut down any heating or cooling systems to minimize the spread of any released fibers.
- Do not disturb the material any more than is needed to take a small sample.
- Place a plastic sheet on the floor below the area that is to be sampled.
- Wet the material using a fine mist of water containing a few drops of detergent before taking the sample. Use of the water/detergent mist will reduce the release of asbestos fibers.
- Carefully cut a piece from the entire depth of the materials using, for example, a small knife, corer, or other sharp object. Place the small piece into a clean container (e.g., a 35mm film canister, small glass or plastic vial, or high-quality resealable plastic bag).
- Tightly seal the container after the sample is in it.
- Carefully dispose of the plastic sheet. Use a damp paper towel to clean up any material on the outside of the container or around the area sampled. Dispose of asbestos materials according to state and local procedures.
- Label the container with an identification number and clearly state when and where the sample was taken.

- Patch the sampled area with the smallest possible piece of duct tape to prevent fiber release.
- Send the sample to an EPA-approved laboratory for analysis. The National Institute for Standards and Technology (NIST) has a list of these laboratories. You can get this list from the Laboratory Accreditation Administration, NIST, Gaithersburg, MD 20899; telephone: 301-975-4016. Your state or local health department may also be able to help.

▶ HOW TO MANAGE AN ASBESTOS PROBLEM

If asbestos is found to be intact and in good condition, it should be left alone; see the list of dos and don'ts in Table 6.1. If the material will be disturbed or it is damaged, you have two viable options. One of these options is the repair of the asbestos. Repair usually involves either sealing or covering asbestos material. Sealing (encapsulation) involves treating the material with

TABLE 6.1 Asbestos Dos and Dont's

DO	DON'T
Keep activities to a minimum in any areas having damaged material that may contain asbestos.	Dust, sweep, or vacuum debris that may contain asbestos.
Take every precaution to avoid damaging asbestos material.	Saw, sand, scrape, or drill holes in asbestos materials.
Have removal and major repair done by people trained and qualified in handling asbestos. It is highly recommended that sampling and minor repair also be done by asbestos professionals.	Use abrasive pads or brushes on power strippers to strip wax from asbestos flooring or its backing. Never use a power stripper on a dry floor.
	Sand or try to level asbestos flooring or its backing. When asbestos flooring needs replacing, install new floor covering over it, if possible.
	Track material that could contain asbestos through the house. If you cannot avoid walking through the area, have it cleaned with a wet mop. If the material is from a damaged area, or if a large area must be cleaned, call an asbestos professional.

a sealant that either binds the asbestos fibers together or coats the material so fibers are not released. Pipe, furnace, and boiler insulation can sometimes be repaired this way. This should be done only by a professional trained to handle asbestos safely.

Covering (enclosure) involves placing something over or around the material that contains asbestos to prevent release of fibers. Exposed insulated piping may be covered with a protective wrap or jacket. With any type of repair, the asbestos remains in place. Repair is usually cheaper than removal, but it may make later removal of asbestos, if necessary, more difficult and costly. Repairs can either be major or minor. The work should be performed by trained, certified professionals.

The other option is removal. You will find that removal is usually the most expensive method and, unless required by state or local regulations, should be the last option considered in most situations. The reason is that removal poses the greatest risk of fiber release. However, removal may be required when remodeling or making major changes in a building that will disturb asbestos material. Also, removal may be called for if asbestos material is damaged extensively and cannot be otherwise repaired. Removal is complex and must be done only by a contractor with special training. Improper removal may actually increase the health risks to you and your family.

Your work as a manager can involve the use of a number of forms and checklists. See Figures 6.1 through 6.13 and Box 6.3 for examples of what you may be required to work with.

▶ ASBESTOS PROFESSIONALS

Asbestos professionals are trained and certified to work with asbestos. Some of the pros are generalists; others are specialists. Asbestos professionals can conduct building inspections, take samples of suspected material, assess its condition, and advise about what corrections are needed and who is qualified to make these corrections. Once again, material in good condition need not be sampled unless it is likely to be disturbed. Professional correction or abatement contractors repair or remove asbestos materials.

Some firms offer combinations of testing, assessment, and correction. A professional hired to assess the need for corrective action should not be connected with an asbestos-correction

	Inspection Report: List of Homogeneous Areas					
	Project Name: _____					
	Address: _____					
	_____					
	_____					
Area #	Area Description	Linear or Sq. Ft.	L S	Friable Y/N	Type S/T/M	ACBM Y/N

Figure 6.1 Inspection report for homogeneous areas.

firm. It is better to use two different firms so there is no conflict of interest. Services vary from one area to another around the country.

The federal government has training courses for asbestos professionals around the country. Some state and local governments also have or require training or certification courses. Ask asbestos professionals to document their completion of federal or state-approved training. Each person performing work on your

PRINCE WILLIAM COUNTY
Department of Development Services – Building Development Division

ASBESTOS RELEASE FORM

Version 2009-01-22

TO: All Applicants for Building Permits for Renovation or Demolition

RE: Asbestos Inspection in Buildings to be Renovated or Demolished

Notice: *State law provides penalties for willful violations of the asbestos laws. The first and second offenses may be prosecuted as class I misdemeanors. A third and any subsequent violations within a three year period are subject to prosecution as class 6 felonies.*

The Virginia Uniform Statewide Building Code requires that a local building department shall not issue a building permit allowing a building for which an initial building permit was issued before January 1, 1985, to be removed or demolished until the local building department receives certification from the owner or his agent that the affected portions of the building have been inspected for the presence of asbestos by an individual licensed to perform such inspections and that no asbestos-containing materials were found or that appropriate response actions will be undertaken in accordance with current requirements. The following form is to be completed by all applicants for Building Permits for renovation or demolition. A completed form will contain one box checked below and must be signed by the owner or authorized agent of the owner(s).

Owner(s) _____

Project Address _____

AS OWNER, OR OWNERS AGENT, OF THE ABOVE BUILDING, I CERTIFY THAT:

☐   The above building was constructed after January 1, 1985.

☐   The above building is a single family dwelling, or is a residential housing building containing four or fewer units, and is exempt from asbestos inspection requirements (Note: This exemption does not apply if the proposed renovation or demolition is for commercial or public development purposes); or

☐   The combined amount of regulated asbestos-containing material involved in the renovation or demolition is less than 260 linear feet on pipes, or less than 160 square feet on other facility components, or less than thirty-five cubic feet off facility components where length or area could not be measured previously, and is exempt from asbestos inspection requirements.

If none of the above boxes have been checked, and if the building permit application is for repair or replacement of roofing, floor covering, or siding materials and the use is not a school, asbestos inspection requirements may be satisfied by checking the following box:

☐   The materials to be repaired or replaced are assumed to contain friable asbestos and that appropriate response actions will be accomplished by a licensed asbestos contractor.

If none of the four boxes above are applicable, check one of the remaining two boxes below.

☐   The affected area of the above building to be renovated or demolished has been inspected for the presence of asbestos by an individual licensed to perform such inspections and that no asbestos-containing materials were found; or

☐   Asbestos-containing materials in the affected area of the above building to be renovated or demolished will be subject to appropriate response actions in accordance with all applicable laws relating to asbestos abatement.

I further certify that the abatement area will not be reoccupied until any required response actions have been completed and final clearances have been measured and found to be within regulated tolerances.

_____   _____

Printed Name of Owner or Owner's Agent   Telephone Number

_____   _____

Signature of Owner or Owner's Agent   Date

Page 1 of 1
Building Development Division. 5 County Complex Court. Prince William. VA. 22192. 703-792-6930. www.pwcgov.org/BDD

Figure 6.2 Asbestos Release Form from Prince William County in Virginia.

job should provide proof of training and licensing in asbestos work, such as completion of EPA-approved training.

State and local health departments or EPA regional offices may have listings of licensed professionals in your area. If you have a problem that requires the services of asbestos professionals,

\# 001189

ASBESTOS SURVEY NOTIFICATION RECORD
Verification of Inspection

AHERA CERTIFIED BUILDING INSPECTOR

1a. Work Site Name, Address, City, Country, State	1b. Owner's Name and Mailing Address Telephone No. ( )
2. Name & Mailing Address of Company or Individual Conducting Asbestos Survey Telephone No.( )	3. Analytical Laboratory Name and Address Telephone No. ( )
4a. Asbestos NESHAP Regulatory Agency Name & Address for Work Site Telephone No.( )	4b. OSHA Regulatory Agency Name and Address for Work Site Telephone No. ( )

5. APPROXIMATE AMOUNT OF ASBESTOS, INCLUDING:	Amount of RACM to be Removed or Generated	Amount of Nonfriable ACM To Be Removed		Amount of Nonfriable ACM Not To Be Removed	
		CAT I	CAT II	CAT I	CAT II
On Facility Components: Pipes (Linear Feet)					
On Facility Components: Surface Area (Sq. Ft.)					
Off Facility Components: Volume (Cubic Feet)					

6. AHERA Building Inspector Certificate No. & Expiration Date

7. Training Provider Name & Phone No.

8a. Number of samples analyzed & date of analysis

8b. ☐ TSI ☐ Ceiling Texture ☐ Duct/Seam Tape ☐ A/C Pipe ☐ A/C Siding/Shingles ☐ VAT/Mastic ☐ Asphaltic Roofing ☐ Add-on Surfacing Texture on wall systems ☐ Other please specify:

9. INSPECTOR'S CERTIFICATION: I hereby declare that the contents of this Asbestos Survey Notification are fully and accurately described above, are classified in all respects to applicable regulations found in title 40, EPA Code of Federal Regulations, Part 61, Subpart M, Asbestos NESHAP, Sec. 61.145(a).

NOTE: The AHERA Building Inspector must retain a copy of this form.

Printed / Typed Name & Title	Signature	MO DAY YR

ASBESTOS ABATEMENT AND DEMOLITION CONTRACTORS

10. Asbestos Removal Contractor / Operator acknowledges receipt of this form Printed / Typed Name, Title, Address & Telephone No.	Signature	MO DAY YR
11. Demolition Contractor / Operator acknowledges receipt of this form. Printed / Typed Name, Title, Address & Telephone No.	Signature	MO DAY YR

BUILDING PERMIT AGENCY

12. Renovation/Demolition Permit Number, date of issuance	Parcel Number	
13. Building Permit Agency acknowledges receipt of this form. Printed / Typed Name & Title	Signature	MO DAY YR

OWNER

Additional copies of this form are distributed to:

- Asbestos Abatement Contractor/Operator
- Demolition Contractor/Operator
- Building Permit Agency
- General Contractor/Subcontractors
- AHERA Building Inspectors/Asbestos Survey Records

Figure 6.3 Asbestos Survey Notification Record.

ASBESTOS INSPECTION FORM

MAINE DEPARTMENT OF ENVIRONMENTAL PROTECTION
Lead & Asbestos Hazard Prevention Program
17 State House Station, Augusta, Maine 04333

This form is used to determine if an inspection for asbestos is required prior to renovation or demolition projects

If your project involves the demolition and or renovation of a single family residence or a residential building with less than 5 units, please answer the following questions to determine whether you need to have your inspection performed by a Maine-licensed Asbestos Inspector:		
Does this demolition/renovation project involve more than ONE residential building at the same site with the same owner?	Y	N
Is this building currently being used, or has it **EVER** been used, as a commercial, government, daycare, office, church, charitable or other non-profit place of business?	Y	N
Is this building to be demolished as part of a highway or road-widening project?	Y	N
Is this building part of a building cooperative, apartment or condo building?	Y	N
Is this building used for military housing?	Y	N
Have other residences or non-residential buildings at this site been scheduled to be demolished now, or in the future as part of a larger project?	Y	N
Is more than ONE building to be lifted from its foundation and relocated?	Y	N
Will this building be intentionally burned for the purpose of demolition or fire department training?	Y	N

IF YOU ANSWER "NO" TO ALL THE QUESTIONS ABOVE, YOUR BUILDING CAN BE INSPECTED BY A KNOWLEDGEABLE NON-LICENSED PERSON AS APPLICABLE (SEE REVERSE SIDE)

ANY "YES" ANSWERS TO THE ABOVE QUESTIONS REQUIRES AN INSPECTION BY A MAINE-LICENSED ASBESTOS INSPECTOR

Remember:

If your renovation project requires the removal of asbestos containing materials, the removal of those materials must be done by a Maine-licensed asbestos abatement contractor.

Before you can demolish any building , including single-family residences, all asbestos materials must be removed from the building. The removal of those materials must be done by a Maine-licensed asbestos abatement contractor, except single-family homeowners may remove some asbestos under certain circumstances (Contact DEP for more information).

With the exception of a single family home, building owners are required to submit the Asbestos Building Demolition Notification to the DEP at least five (5) working days prior to the demolition **EVEN IF NO ASBESTOS** is present

Once the asbestos is removed, the renovation or demolition project may be performed by any preferred contractor.

I CERTIFY THAT THE ABOVE INFORMATION IS CORRECT		
Print Name: Owner/Agent	Title	Signature
Telephone #	FAX #	Date

Keep this completed form for your records

Figure 6.4 Asbestos Inspection Form from the Maine Department of Environmental Protection.

Sample Recordkeeping Form

Form 1. A sample form for recording information during ACM reassessment.

Reinspection of Asbestos-Containing Materials

Location of asbestos-containing material (address, building, room, or general description]

_____

_____

_____

_____

Type of asbestos-containing material(s):

1. Sprayed-or troweled-on ceilings or walls
2. Sprayed-or troweled-on structural members
3. Insulation on pipes, tanks, or boiler
4. Other (describe)

_____

_____

Abatement Status:

1. The material has been encapsulated_____ , enclosed_____, neither_____, removed_____ .

Assessment:

1. Evidence of physical damage _____

_____

2. Evidence of water damage: _____

_____

3. Evidence of delamination or other damage: _____

_____

4. Degree of accessibility of the material: _____

_____

5. Degree of activity near the material: _____

_____

6. Location in an air plenum, air shaft, or airstream: _____

_____

7. Other observations (including the condition of the encapsulant or enclosure, if any): _____

_____

*Recommended Action: _____

_____

Signed:_____   Date: _____
         (evaluator)

Figure 6.5 Sample recordkeeping form.

Fiber Release Episode Report

1. Address, building, and room number(s) (or description of area) where episode occurred:

   _____

   _____

   _____

   _____

2. The release episode was reported by _____ _____

   on _____ (date).

3. Describe the episode: _____

   _____

   _____

   _____

   _____

   _____

   _____

   _____

4. The asbestos-containing material was ____/was not ____ cleaned up according to approved procedures.

5. Describe the cleanup: _____

   _____

   _____

   _____

   _____

Signed: _____ Date: _____
          (Asbestos Program Manager)

Figure 6.6 Fiber Release Episode Report.

check their credentials carefully. Hire professionals who are trained, experienced, reputable, and accredited, especially if accreditation is required by state or local laws.

Before hiring a professional, ask for references from previous clients. Find out if they were satisfied. Ask whether the professional has handled similar situations. Get cost estimates from several professionals because the charges for these services can vary. There are numerous recommendations for

---

Inspection Report Compliance Checklist

This checklist is designed to enable you to determine if the inspection report is complete and contains each and every element required by law.

GENERAL:

_____ 1. The date of the inspection
_____ 2. The signature of each accredited person making the inspection
_____ 3. The State of accreditation of each accredited person making the inspection
_____ 4. If applicable, the accreditation number of each accredited person making the inspection

INVENTORY OF LOCATIONS:

_____ 5. An inventory of the locations of the homogeneous areas where samples were collected
_____ 6. The exact location where each bulk sample was collected
_____ 7. The date(s) that each sample was collected
_____ 8. The homogeneous areas where friable suspected ACBM is assumed to be ACBM
_____ 9. The homogeneous areas where nonfriable suspected ACBM is assumed to be ACBM

SAMPLING:

_____ 10. A description of the manner used to determine sampling locations
_____ 11. The name and signature of each accredited inspector who collected the samples
_____ 12. The state of accreditation of each accredited inspector who collected the samples
_____ 13. If applicable, the accreditation number of each accredited inspector who collected the samples

MATERIALS IDENTIFIED IN HOMOGENEOUS AREAS:

_____ 14. A list of whether the homogeneous areas identified are surfacing material, thermal system insulation, or miscellaneous material

ASSESSMENTS:

_____ 15. Assessments made of friable material
_____ 16. The name and signature of each accredited inspector who made the assessment
_____ 17. The state of accreditation of each accredited inspector who made the assessment
_____ 18. If applicable, the accreditation number of each accredited inspector who made the assessment

---

Figure 6.7 Inspection Report Compliance Checklist.

managers of asbestos projects. You can see examples of such recommendations in Figures 6.14 through 6.20 and Boxes 6.4 through 6.9.

Even though private homes are usually not covered by the asbestos regulations that apply to schools and public buildings,

Checklist of Final Air Clearance Documentation

This checklist will indicate whether each final clearance was properly documented.

___ 1. The name and signature of any person collecting any air sample required to be collected at the completion of a response action

___ 2. The locations where those samples were collected

___ 3. The name and address of the laboratory, analyzing the samples

___ 4. The date(s) of analysis

___ 5. The results of analysis

___ 6. The method of analysis

___ 7. The name and signature of the person performing the analysis

___ 8. Evidence that the laboratory is NVLAP accredited

Figure 6.8 Checklist of Final Air Clearance Documentation.

professionals should still use procedures described during federal or state-approved training. General contractors should be alert to the chance of misleading claims by asbestos consultants and contractors.

There have been reports of firms incorrectly claiming that asbestos materials in buildings must be replaced. In other cases, firms have encouraged unnecessary removals or performed them improperly. Unnecessary removals are a waste of money. Improper removals may actually increase the health risks to anyone in or around the affected building. To guard against this problem, know what services are available and what procedures and precautions are needed to do the job properly.

In addition to general asbestos contractors, you may select a roofing, flooring, or plumbing contractor trained to handle asbestos when it is necessary to remove and replace roofing, flooring, siding, or asbestos-cement pipe that is part of a water system. Normally, roofing and flooring contractors are exempt from state and local licensing requirements because they do not perform any other asbestos correction work. Call 1-800-USA-ROOF for names of qualified roofing contractors in your area (Illinois residents, call 708-318-6722).

DETACHED HOME ASBESTOS SURVEY CHECKLIST*

	Suspect Materials	Quantity (SF or LF)	# of Samples	Number Positive
☐	Resilient Floor Covering (Linoleum, etc.)	___	___	___
☐	Floor Tile (9"×9" or 12"×12" nonceramic)	___	___	___
☐	Dry Wall with Joint Compound	___	___	___
☐	Skim Coat (Plaster)	___	___	___
☐	Roof Shingles	___	___	___
☐	Roofing Felt Paper	___	___	___
☐	Roof Flashing	___	___	___
☐	Penetration Mastics	___	___	___
☐	Ceiling Tiles (9"×9", 2'×2', or 4'×2')	___	___	___
☐	Textured Ceiling Materials (Stucco, popcorn, etc.)	___	___	___
☐	Duct Work Insulation Material	___	___	___
☐	Duct Tape Attached to Duct Work	___	___	___
☐	Attic or "Blown" Insulation	___	___	___
☐	Pipe Insulation	___	___	___
☐	Insulation on Pipe Elbows or Pipe Connections	___	___	___
☐	Caulking around Doors and Windows	___	___	___
☐	Window Glazing	___	___	___
☐	Exterior Siding (Cementitious) (a.k.a. *Transite*)	___	___	___
☐	Undercoating Material Attached to Underside of Stainless Steel Sinks	___	___	___
☐	Heat Deflector Attached Behind Wall or Ceiling-Mounted Lights	___	___	___
	Miscellaneous Materials:			
☐	_____	___	___	___
☐	_____	___	___	___
☐	_____	___	___	___

***NOTE: For Asbestos Inspections of Residential-Type Houses.**
To be used as a guideline to achieve substantial compliance with the "thorough inspection" provisions of state and federal asbestos rules and regulations.

Figure 6.9 Detached Home Asbestos Survey Checklist.

For information on asbestos in floors, read *Recommended Work Procedures for Resilient Floor Covers*. You can write for a copy from the Resilient Floor Covering Institute, 966 Hungerford Drive, Suite 12-B, Rockville, MD 20850. Enclose a stamped business-size, self-addressed envelope.

STATE OF DELAWARE
ASBESTOS INSPECTION FORM

FACILITY NAME:				
ADDRESS:			TAX PARCEL:	
CITY:	COUNTY:	STATE:	ZIP:	
SITE CONTACT NAME:		CONTACT PHONE:		
OWNER NAME:				
OWNER ADDRESS:				
CITY:	COUNTY:	STATE:	ZIP:	
OWNER CONTACT:		OWNER PHONE:		

FACILITY DESCRIPTION: G Agricultural　G Commercial　G Industrial　G Institutional　G Public　G Residential

BUILDING DESCRIPTION (describe structure and size): _____

NOTE: This Survey form was designed to be used for ONE Building/Structure only. Use additional forms for additional structures.

PROFESSIONAL SERVICE FIRM:				
ADDRESS:				
CITY:	COUNTY:	STATE:	ZIP:	
INSPECTOR'S NAME:		PHONE NUMBER:		
INSPECTION'S CERTIFICATION: PM# –		PROFESSIONAL SERVICE FIRM CERTIFICATION: PS# –		

TYPE OF INSPECTION:　G RENOVATION　G DEMOLITION　｜　DATE OF INSPECTION:

IS ASVESTOS CONTAINING MATERIAL PRESENT?　　G YES　G　NO　*See summary results on page 2.*

I hereby certify, that I am a Delaware Licensed inspector employed by a Delaware Licensed Professional Service Firm and that the building and/or contents therein located at the property identified above have been inspected for asbestos containing materials in accordance with the State of Delaware Regulations Governing the Control of Air Pollution, Regulation #21 Section 10

Name_____　　　　　Title: _____

Signature_____　　　　Date: _____

If ACM is present and will be disturbed, removed or abated:
Name of Abatement Company (Print Company Name) _____
Phone # _____ Asbestos Abatement Contractor License # _____
The State of Delaware requires a licensed asbestos abatement contractor for all abatement projects except for work performed in an owner-occupied single family dwelling.

Asbestos Abatement & Demolition/Renovation Notification Form submitted to DNREC/USEPA

Region 3 on _____ (insert date) _____ (insert DOANS #)

The Notification must be submitted a minimum of 10 days prior to beginning the abatement project (see 40 CFR 61 Subpart M).

AQM-ASB-001　　Rev. 3　　Dec 2006　　　　　　　　　　　　　　　　　　Page 1 of 2

Figure 6.10 Asbestos Inspection Form from the State of Delaware.

If You Hire a Professional Asbestos Inspector

If you hire a professional asbestos inspector, there are some steps that you should follow, including:

- Make sure that the inspection will include a complete visual examination and the careful collection and lab analysis of samples. If asbestos is present, the inspector should provide a written evaluation that describes its location and the extent of

| SUMMARY OF ABESTOS SURVEY / INSPECTION | | | | | |
Material/Product Surveyed[1]	Sampled? Yes/No[2]	ACM Present (%)[3]	Condition of ACM/ Suspected ACM	No ACM Present T	Abatement Required? Yes/No
ROOFING & SIDING					
o Roof felt shingles	❑ Yes ❑ No				❑ Yes ❑ No
o Roofing shingles	❑ Yes ❑ No				❑ Yes ❑ No
o Roofing tiles	❑ Yes ❑ No				❑ Yes ❑ No
o Siding shingles	❑ Yes ❑ No				❑ Yes ❑ No
o Clapboards	❑ Yes ❑ No				❑ Yes ❑ No
o Other	❑ Yes ❑ No				❑ Yes ❑ No
WALLS & CEILINGS					
o Ceiling tiles	❑ Yes ❑ No				❑ Yes ❑ No
o Ceiling tile mastic	❑ Yes ❑ No				❑ Yes ❑ No
o Sprayed/troweled coating	❑ Yes ❑ No				❑ Yes ❑ No
o Asbestos-cement sheet	❑ Yes ❑ No				❑ Yes ❑ No
o Paneling, tile, baseboard mastic	❑ Yes ❑ No				❑ Yes ❑ No
o Spackle/joint compounds	❑ Yes ❑ No				❑ Yes ❑ No
o Textured paints	❑ Yes ❑ No				❑ Yes ❑ No
o Millboard, rollboard	❑ Yes ❑ No				❑ Yes ❑ No
o Vinyl wallpaper	❑ Yes ❑ No				❑ Yes ❑ No
o Insulation board	❑ Yes ❑ No				❑ Yes ❑ No
o Other	❑ Yes ❑ No				❑ Yes ❑ No
FLOORS					
o Vinyl-asbestos tile	❑ Yes ❑ No				❑ Yes ❑ No
o Asphalt-asbestos tile	❑ Yes ❑ No				❑ Yes ❑ No
o Resilient sheet flooring	❑ Yes ❑ No				❑ Yes ❑ No
o Mastic adhesives	❑ Yes ❑ No				❑ Yes ❑ No
o Other	❑ Yes ❑ No				❑ Yes ❑ No
PIPES & BOILERS					
o Cement pipe and fittings	❑ Yes ❑ No				❑ Yes ❑ No
o Block insulation	❑ Yes ❑ No				❑ Yes ❑ No
o Preformed pipe wrap	❑ Yes ❑ No				❑ Yes ❑ No
o Corrugated asbestos paper	❑ Yes ❑ No				❑ Yes ❑ No
o Paper tape	❑ Yes ❑ No				❑ Yes ❑ No
o Putty (mudding)	❑ Yes ❑ No				❑ Yes ❑ No
o Other	❑ Yes ❑ No				❑ Yes ❑ No
OTHER PRODUCTS					
o Window glazing putty	❑ Yes ❑ No				❑ Yes ❑ No
o Building caulk	❑ Yes ❑ No				❑ Yes ❑ No
o Gaskets/packing	❑ Yes ❑ No				❑ Yes ❑ No
o Clothing/cloth/blankets	❑ Yes ❑ No				❑ Yes ❑ No
o Cement/mortar	❑ Yes ❑ No				❑ Yes ❑ No
o Metal-clad firebrick	❑ Yes ❑ No				❑ Yes ❑ No
o Gunnite/fire-proofing spray	❑ Yes ❑ No				❑ Yes ❑ No
o Hot-tops (ingot mold covers and inserts)	❑ Yes ❑ No				❑ Yes ❑ No
	❑ Yes ❑ No				❑ Yes ❑ No

[1]This list is not an exclusive list of potential materials containing asbestos and the inspector should use it only as a minimal reference of potential asbestos containing materials present.
[2]No sampling is required if the inspector suspects that the materials are ACM and treats them as ACM. For a suspect material to be classified as non-ACM, a minimum number of samples must be collected and analyzed as required by AHERA/ASHARA regulations.
[3]All materials identified as having an asbestos content greater than 1% are considered to be regulated asbestos containing materials (RACM).
NOTE: If this structure is to be demolished by intentional burning by a Delaware Fire Company. This form must be provided to the Fire Company in order for DNREC-AQM to process the Fire Fighting Instruction by intentional burning application.

AQM-ASB-001   Rev. 3   Dec 2006   Page 2 of 2

Figure 6.10 *Cont'd.*

damage, and then give recommendations for correction or prevention.
• Make sure an inspecting firm makes frequent site visits if it is hired to assure that a contractor follows proper procedures and requirements. The inspector may recommend and perform checks after the correction to assure the area has been properly cleaned. This is always a good idea.

Bulk Sampling Requirements	
Type of Material	**Samples Required**
Friable Surfacing Material:	
• Area ≤ 1,000 sq. ft.	3
• Area > 1,000 sq. ft. but ≤ 5,000 sq. ft.	5
• Area > 5,000 sq. ft.	7
Thermal System Insulation (TSI):	
• TSI not assumed to be ACBM	3
• Patched TSI not assumed to be ACBM (if patched section < 6 linear or square ft.)	1
• Each insulated mechanical system not assumed to be ACBM where cement or plaster is used on fittings such as tees, elbows, or valves	Samples in a manner sufficient to determine if material is or is not ACBM*
Friable Miscellaneous Material Not Assumed to Be ACBM	Samples in a manner sufficient to determine if material is or is not ACBM*
Nonfriable Suspected ACBM Not Assumed to Be ACBM	Samples in a manner sufficient to determine if material is or is not ACBM*

* EPA recommends that three samples be taken to meet this requirement.

Note: The designation of ACM for a homogeneous area based on one positive bulk sample result is acceptable.

Figure 6.11 Bulk Sampling Requirements.

If You Hire a Corrective-Action Contractor

If you hire a corrective-action contractor, you should consider the following:

- Check with your local air pollution control board, the local agency responsible for worker safety, and the Better Business Bureau. Ask whether the firm has had any safety violations. Find out whether there are legal actions filed against it.
- Insist that the contractor use the proper equipment to do the job. The workers must wear approved respirators, gloves, and other protective clothing.
- Before work begins, get a written contract specifying the work plan; cleanup; and the applicable federal, state, and local regulations that the contractor must follow (such as notification requirements and asbestos disposal procedures). Contact your state and local health departments, the EPA's regional office, and the OSHA's regional office to find out what the regulations are. Be sure the contractor follows local

Functional Spaces/Homogeneous Areas	
Building: _____ _____ _____ _____	
Functional Space Letter	Homogeneous Areas by Number (Obtained from Form 1)
Key: L/S = Linear Feet/Square Feet; S/T/M = Surfacing/Thermal/Miscellaneous	

Figure 6.12 Functional Spaces/Homogeneous Areas.

asbestos removal and disposal laws. At the end of the job, get written assurance from the contractor that all procedures have been followed.

- Assure that the contractor avoids spreading or tracking asbestos dust into other areas of your home. The contractor should seal the work area from the rest of the house using plastic sheeting and duct tape, and also turn off the heating and air conditioning system. For some repairs, such as pipe insulation removal, plastic glove bags may be adequate. They must be sealed with tape and properly disposed of when the job is complete.

Bulk Sample Log	
School_____Date Sampled_____	
Homogeneous Area_____Sampler's Name_____	
Functional Space/Room_____Accreditation No._____	
Linear Feet_____ Type of Suspect Material_____	
Square Feet_____ Surfacing TSI Misc.	
_____ Friable _____ Nonfriable	
Manner of Sampling_____	
AREA DESCRIPTION_____	

Number	Location

Figure 6.13 Bulk Sample Log.

- Make sure the work site is clearly marked as a hazard area. Do not allow unintended people or pets into a work area until work is completed.
- Insist that the contractor apply a wetting agent to the asbestos material with a hand sprayer that creates a fine mist before removal. Wet fibers do not float in the air as easily as dry fibers and will be easier to clean up.

> ### Box 6.3. Key Points about Recordkeeping
>
> Each **LEA** must **maintain a copy of its management plan** in its administrative office, and the plan must be available to persons for inspection without cost or restriction.
>
> Each **school** must **maintain a copy of the management plan** for that school in its administrative office, and the plan must be available to persons for inspection without cost or restriction.
>
> The LEA must also maintain records of events that occur after submission of the management plan; these records include training information, periodic surveillance information, cleaning information, small-scale, short-duration O&M activity information, information on O&M activities other than small-scale, short-duration, information on fiber release episodes, information on response actions and preventive measures, and air sampling information. These records should be included in the management plans in a timely manner.
>
> For each homogeneous area where all ACBM has been removed, the LEA must retain the records of events for **3 years** after the next reinspection, or for an equivalent period.
>
> It is the responsibility of the LEA designated person to ensure that complete and up-to-date records are maintained and included in the management plans.

- Make sure the contractor does not break removed material into small pieces. This could release asbestos fibers into the air. Pipe insulation was usually installed in preformed blocks and should be removed in complete pieces.
- Upon completion, ensure that the contractor cleans the area well with wet mops, wet rags, sponges, or high-efficiency particulate air (HEPA) vacuum cleaners. A regular vacuum cleaner must never be used. Wetting helps reduce the chance of spreading asbestos fibers in the air. All asbestos materials and disposable equipment and clothing used in the job must be placed in sealed, leakproof, and labeled plastic bags.

The work site should be visually free of dust and debris. Air monitoring (to make sure there is no increase of asbestos fibers in the air) may be necessary to assure that the contractor's job is done properly. This should be done be someone not connected with the contractor.

	AHERA Asbestos Management Plan Self-Audit Checklist for Designated Persons	
School:		**Phone:**
Address:		
County:		
Local Education Agency:		**Phone:**
Address:		
Designated Person:		**Phone:**
Address:		
Date Checklist Completed by Designated Person:		
Designated Person's Signature:		
Yes No N/A N/A - Not Applicable	**School:**	
	General Information	
☐ ☐ ☐	1. Has an Asbestos Management Plan been developed for your school? (40 CFR § 763.9:	
☐ ☐ ☐	2. Does the Local Education Agency (LEA) have a complete and up-to-date copy of the school management plan in both the LEA's administrative office and the school's administrative office (40 CFR § 763.93(g)(2)-(3	
☐ ☐ ☐	3. Was the management plan developed by an accredited management planner?	**Did you know?** Your LEA *may* require each management plan to contain a statement signed by an accredited management plan developer that he/she has prepared or assisted in the preparation of the plan or has reviewed the plan and that the plan is in compliance with 40 CFR 763, Subpart E. The management plan developer that signs the statement may not also implement the plan (40 CFR § 763.93(f)). (40 CFR § 763.93(e

Figure 6.14 The EPA's Self-Audit Checklist for Designated Persons.

Do not dust, sweep, or vacuum debris that may contain asbestos. These steps will disturb tiny asbestos fibers and may release them into the air. Remove dust by wet mopping or have trained asbestos contractors use a special HEPA vacuum cleaner.

As a managing general contractor, you should monitor the work of your asbestos professionals. This does not mean entering the work site during abatement. If you do this, make sure that you are equipped with suitable protective personal

Yes	No	N/A	School:

N/A - Not Applicable

Yes	No	N/A	
☐	☐	☐	4. For each consultant who contributed to the management plan, does the plan include the following: • Consultant's name? • A statement that he/she is accredited under the state accreditation program or another state's accreditation program or an EPA-approved course? (40 CFR § 763.93 (e)(12)(i)-(ii)) Note: Although not required, EPA suggests including in the AMP the name of the training agency, the course name and date, and a copy of the accreditation certificate for each consultant. *Tip: See suggested Model AMP Form 1 - Contact Information
☐	☐	☐	5. Does the management plan include a list of the name and address of each building used as a school building and identify whether the school building has: • Friable ACBM (asbestos-containing building material)? • Nonfriable ACBM? • Friable and nonfriable suspected ACBM assumed to be ACM (asbestos-containing material)? (40 CFR §§ 763.93(a)(1)-(2) and 763.93(e)(1)) *Tip: See Model AMP Form 2—School Building List.
☐	☐	☐	6. If a new school building was constructed after October 12, 1988 and is asbestos-free, does the management plan include the following and has a copy of same been provided by the LEA to the EPA Regional Office: • A statement signed by an architect or project engineer responsible for the construction of the building, or by an accredited inspector, indicating that no ACBM was specified as a building material in any construction document for the building, or, to the best of his or her knowledge, no ACBM was used as a building material in the building? (40 CFR § 763.99(a)(7)) *Tip: See Model AMP Form 2 - School Building List
☐	☐	☐	7. Does the management plan include a copy of any of the statements required under 40 CFR § 763.99(a)(1)–(7) to support an exclusion from inspection that the school may qualify for under 40 CFR § 763.99 and has a copy of any such statement been provided by the LEA to the Regional Office? (40 CFR § 763.99) Note: The exclusion under 40 CFR § 763.99(a)(7) is also covered under Checklist question number 6.

Figure 6.14 *Cont'd*

Yes No N/A	School:
N/A - Not Applicable	
☐ ☐ ☐	8. Does the management plan include the following information about the LEA Designated Person (DP): • Name, address, and telephone number of the DP? • Course name, dates, and hours of training that the DP attended to carry out his or her AHERA duties? • Signed statement by the DP that the LEA's general responsibilities under 40 CFR § 763.84 have been or will be met? (40 CFR § 763.93(e)(4) and (i Note: Although not required, EPA suggests including in the AMP the name of the training agency and a copy of the DP's training certificates. *Tip: See Model AMP Form 1 - Contact Information and Form 3 - Designated Person Assurances
☐ ☐ ☐	9. Does the management plan include the following recommendations: • A plan for reinspection required under 40 CFR § 763.85? • A plan for operations and maintenance activities (including initial cleaning) required under 40 CFR § 763.91? • A plan for periodic surveillance required under 40 CFR § 763.92? • A description of the management planner's recommendation for additional cleaning under 40 CFR § 763.91 (c)(2), as part of an operations and maintenance program, and the response of the LEA to that recommendation? (40 CFR § 763.93(e)(9 *Tip: See Model AMP Form 10 - Plan for Reinspection, Form 14 - Plan for Operations and Maintenance Activities, Form 18 - Periodic Surveillance Plan/Report, and Form 16 Cleaning Record
☐ ☐ ☐	10. Does the management plan include an evaluation of resources needed to carry out response actions, reinspections, operations and maintenance, and periodic surveillance and training? (40 CFR § 763.93(e)(11) *Tip: See suggested Model AMP Form 4 - Evaluation of Resources
☐ ☐ ☐	11. Does the management plan include a record of the minimum 2 hours of awareness training required under 40 CFR § 763.92(a)(1) for all maintenance and custodial staff who may work in a building that contains ACBM, whether or not they are required to work with ACBM, and does the record include the following information: • Person's name and job title? • Date training was completed? • Location of training? • Number of hours completed? **Did you know?** New custodial and maintenance employees must be trained within 60 days after starting work (40 CFR §763.92(a)(1)). (40 CFR §§ 763.93(h) and 763.94(Note: Although not required, EPA suggests including in the AMP the name of the training agency, th course name, and a copy of the accreditation certificate for each staff person. *Tip: See Model AMP Form 5 - Training Record for Maintenance and Custodial Staff

Figure 6.14 *Cont'd*

Yes No N/A	School:
N/A - Not Applicable	
☐ ☐ ☐	12. Does the management plan include a record of the additional 14 hours of training required under 40 CFR § 763. 92(a)(2) for maintenance and custodial staff who conduct any activities that will result in the disturbance of ACBM and does the record include the following information: • Person's name and job title? • Date training was completed? • Location of training? • Number of hours completed? (40 CFR §§ 763.93(h) and 763.94(c)) Note: Although not required, EPA suggests including in the AMP the name of the training agency, the course name, and a copy of the accreditation certificate for each staff person. *Tip: See Model AMP Form 5 - Training Record for Maintenance and Custodial Staff
	Inspections and Reinspections
☐ ☐ ☐	13. For inspections conducted before 12/14/87 (i.e., the effective date of the 10/30/87 EPA Asbestos-Containing Materials in Schools Rule), does the management plan include the following information: • Date of inspection? • Blueprint, diagram, or written description of each school building that identifies clearly each location and approximate square or linear footage of homogenous/sampling area sampled for ACM? • If possible, the exact locations where the bulk samples were collected and the dates of collection? • A copy of the analyses of any bulk samples, dates of analyses, and a copy of any other laboratory reports pertaining to the analyses. • Description of response actions or preventive measures taken, including, if possible, the names and addresses of all contractors, start and completion dates and air clearance sample results? • Description of assessments of material identified prior to 12/14/87 as friable ACBM or friable suspected ACBM assumed to be ACM, and the name, signature, state of accreditation and if, applicable, the accreditation number of the person making the assessments (i.e., inspector)? (40 CFR § 763.93(e)(2)(i)-(v)) *Tip: See Model AMP Form 6 - Inspection Cover Sheet, Form 8 - Homogeneous Area/Bulk Sample Summary, Form 9 - Homogeneous Area/Bulk Sample Diagram, Form 12 - Implementation of Response Actions, and Form 7 - Room/Functional Space Assessment
☐ ☐ ☐	14. Does the management plan include for each inspection and reinspection conducted under 40 CFR § 763.85 the following information: • Date of the inspection or reinspection? • Name, signature, state of accreditation, and, if applicable, the accreditation number for each accredited inspector performing the inspection or reinspection? (40 CFR § 763.93(e)(3)(i)) Note: Although not required, EPA suggests including in the AMP the name of the training agency, the course name and date, and a copy of the accreditation certificate for each inspector. *Tip: See Model AMP Form 6 - Inspection Cover Sheet

Figure 6.14 *Cont'd*

Yes No N/A	School:
N/A - Not Applicable	
☐ ☐ ☐	15. Does the management plan include for each inspection and reinspection conducted under 40 CFR § 763.85 the following sampling information: • Blueprint, diagram, or written description of each school building that identifies clearly each location and approximate square or linear footage of homogeneous areas where material was sampled for ACM? • Exact location where each bulk sample was collected and the date of collection of each bulk sample? • Homogeneous areas where friable suspected ACBM is assumed to be ACM? • Homogeneous areas where nonfriable suspected ACBM is assumed to be ACM? • Description of the manner used to determine sampling locations? • The name, signature, state of accreditation, and, if applicable, the accreditation number for each accredited inspector who collected samples? (40 CFR § 763.93 (e)(3)(ii)-(iii) Note: For details on how to collect bulk samples, see 40 CFR § 763.86. Although not required, EPA suggests including in the AMP the name of the training agency, the course name and date, and a copy of the accreditation certificate for each inspector who collected the samples. *Tip: See Model AMP Form 6 - Inspection Cover Sheet, Form 8 - Homogeneous Area/Bulk Sample Summary, and Form 9 - Homogeneous Area/Bulk Sample Diagram
☐ ☐ ☐	16. Does the management plan include for each inspection and reinspection conducted under 40 CFR § 763.85 the following information on the analysis of the bulk samples and has it been submitted to the DP for inclusion in the plan within 30 days of the analysis: • Copy of the analysis of any bulk samples collected and analyzed? • Name and address of any laboratory that analyzed bulk samples? • A statement that any laboratory used meets the applicable laboratory accreditation requirements of 40 CFR § 763.87(a)? • Dates of any analyses performed? • Name and signature of the person performing each analysis? (40 CFR §§ 763.87(d) and 763.93(e)(3)(iv) Note: For details on how to submit bulk samples for analysis, see 40 CFR § 763.87.
☐ ☐ ☐	17. Does the management plan include for each inspection and reinspection conducted under 40 CFR § 763.85 the following assessment information and has it been submitted to the DP for inclusion in the plan within 30 days of the assessment: • Written assessments (signed and dated) required to be made under 40 CFR § 763.88 of all ACBM and suspected ACBM assumed to be ACBM? • Name, signature, state of accreditation, and, if applicable, the accreditation number of each accredited person making the assessment (i.e., inspector(s)) (40 CFR §§ 763.88(a)(2) and 763.93(e)(3)(v) Note: Although not required, EPA suggests including in the AMP the name of the training agency, the course name and date, and a copy of the accreditation certificate for each inspector making the assessment. *Tip: See Model AMP Form 6 - Inspection Cover Sheet and Form 7 - Room/Functional Space Assessment

Figure 6.14 *Cont'd*

Yes No N/A	School:
N/A - Not Applicable	
☐ ☐ ☐	18. Has the following information about the inspection been recorded and submitted to the DP for inclusion in the management plan within 30 days of the inspection: • Inspection report with the date of inspection signed by each accredited inspector making the inspection, the state of accreditation, and if applicable, his/her accreditation number? • Inventory of the locations of the homogeneous areas where samples are collected, exact location where each bulk sample is collected, dates that samples are collected, homogeneous areas where friable suspected ACBM is assumed to be ACM and homogeneous areas where nonfriable suspected ACBM is assumed to be ACM? • Description of the manner used to determine sampling locations, the name and signature of each accredited inspector who collected the samples, state of accreditation, and, if applicable, his or her accreditation number? • List of whether the homogeneous areas identified under 40 CFR § 763.85(a)(4)(vi)(B) of this section are surfacing material, thermal system insulation, or miscellaneous material? • Assessments of friable material (signed and dated), the name and signature of each accredited inspector making the assessment, state of accreditation, and if applicable, his or her accreditation number? (40 CFR §§ 763.85(a)(4)(vi)(A)-(E) and 763.88(a)(2)) Note: For further details on activities conducted during inspection (e.g., visually inspect/touch material), see 40 CFR § 763.85(a)(4)(i)-(v) *Tip: See Model AMP Form 6 - Inspection Cover Sheet, Form 7 - Room/Functional Space Assessment, Form 8 - Homogeneous Area/Bulk Sample Summary and Form 9 - Homogeneous Area/Bulk Sample Diagram
☐ ☐ ☐	19. Has the following information about the reinspection been recorded and submitted to the DP for inclusion in the management plan within 30 days of the reinspection: • Date of reinspection, name and signature of the person making the reinspection, state of accreditation, and if applicable, his or her accreditation number, and any changes in the condition of known or assumed ACBM? • Exact location where samples were collected during the reinspection, a description of the manner used to determine sampling locations, the name and signature of each accredited inspector who collected the samples, state of accreditation, and, if applicable, his or her accreditation number? • Any assessments or reassessments of friable material, date of the assessment or reassessment, the name and the signature of the accredited inspector making the assessments, state of accreditation, and if applicable, his or her accreditation number? (40 CFR §§ 763.85(b)(3)(vii)(A)-(C) and 763.88(a)(2)) Note: At least once every 3 years after a management plan has been in effect, a reinspection must be conducted by an accredited inspector of all friable and nonfriable known or assumed ACBM in each school building that the LEA leases, owns, or otherwise uses as a school building (40 CFR § 763.85(b)(1)-(2)). For further details on activities conducted during a reinspection (e.g., visually reinspect/touch material), see 40 CFR § 763.85(b)(3)(i)-(vi). *Tip: See Model AMP Form 6 - Inspection Cover Sheet, Form 7 - Room/Functional Space Assessment, Form 8 - Homogeneous Area/Bulk Sample Summary, Form 9 - Homogeneous Area/Bulk Sample Diagram

Figure 6.14 *Cont'd*

Yes No N/A	School:
N/A - Not Applicable	

			Response Actions
☐	☐	☐	20. Does the management plan include the recommedations made to the LEA regarding response actions under 40 CFR § 763.88(d) and the following information about the accredited management planner: • Name, signature, state of accreditation, and, if applicable, the accreditation number for each accredited management planner making the recommendations? (40 CFR §§ 763.88(d) and 763.93(e)(5)) Note: Although not required, EPA suggests including in the AMP the name of the training agency, the course name and date, and a copy of the accreditation certificate for each accredited person making the recommendations. *Tip: See Model AMP Form 11 - Recommended Response Actions
☐	☐	☐	21. Does the management plan include a detailed description of preventive measures and response actions to be taken, including the following: **Did you know?** The LEA may select, from the response actions that protect human health and the environment, the least burdensome action (40 CFR § 763.90(a)). • Methods to be used for any friable ACBM? • Locations where such measures and actions will be taken? • Reasons for selecting the response action or preventive measure? • Schedule for beginning and completing each preventive measure or response action? (40 CFR § 763.93(e)(6)) Note: For further details on how to conduct response actions, see 40 CFR § 763.90 *Tip: See Model AMP Form 11 - Recommended Response Actions
☐	☐	☐	22. Does the management plan include one of the following statements for the person or persons who inspected for ACBM and who will design or carry out response actions, except for operations and maintenance, with respect to the ACBM: • Statement that he/she is accredited under the state accreditation program, or that the LEA has used (or will use) persons accredited under another state's accreditation program or an EPA-approved course? (40 CFR § 763.93(e)(7)) *Tip: See note on Model AMP Form 3 - Designated Persons Assurances

Figure 6.14 *Cont'd*

Yes	No	N/A	School:
N/A - Not Applicable			

Yes	No	N/A	
☐	☐	☐	23. Does the management plan include a detailed written description of each preventive measure and response action taken for friable and nonfriable ACBM and friable and nonfriable suspected ACBM assumed to be ACM, including the following: • Methods used? • Location where the measure or action was taken? • Reasons for selecting the measure or action? • Start and completion dates of the work? • Names and addresses of all contractors involved and, if applicable, their state of accreditation and accreditation numbers? • If ACBM is removed, the name and location of storage or disposal site of the ACM? *(40 CFR § 763.94(b)(1))* Note: Although not required, EPA suggests including in the AMP a copy of the accreditation. *Tip: See Model AMP Form 12 - Implementation of Response Actions
☐	☐	☐	24. Does the management plan include the following sampling information required to be collected at the completion of certain response actions specified by 40 CFR § 763.90(i): • Name and signature of any person collecting any air sample required to be collected? • Locations where samples were collected? • Date of collection? • Name and address of the laboratory analyzing the samples? • Date of analysis? • Results of analysis? • Method of analysis? • Name and signature of the person performing the analysis? • Statement that the laboratory meets the applicable laboratory accreditation requirements of 40 CFR § 763.90(i)(2)(ii)? *(40 CFR § 763.94(b)(2))* *Tip: See Model AMP Form 12 - Implementation of Response Actions
☐	☐	☐	25. Does the management plan include a detailed description in the form of a blueprint, diagram, or written description of any ACBM or suspected ACBM assumed to be ACM that remains in the school once response actions are undertaken under 40 CFR § 763.90, and is the description updated as response actions are completed? *(40 CFR § 763.93(e)(8))*
☐	☐	☐	26. For each homogeneous area where all ACBM has been removed, have records been retained in the management plan for at least 3 years after the next reinspection required under 40 CFR § 763.85(b)(1), or for an equivalent period? **Did you know?** Significantly damaged friable surfacing ACM or significantly damaged friable miscellaneous ACM must be immediately isolated and access must be restricted unless isolation is not necessary to protect human health and the environment. Then, this material must be removed, or depending on whether enclosure or encapsulation would be sufficient to protect human health and the environment, enclosed or encapsulated (40 CFR § 763.90(d)(1) - (2)). *(40 CFR §§ 763.93(h) and 763.94(a)*

igure 6.14 *Cont'd*

Yes No N/A N/A - Not Applicable	**School:**
	Operations and Maintenance
☐ ☐ ☐	27. Does the management plan include a record of each cleaning conducted under 40 CFR § 763.91 (c), including the following: • Name of each person performing the cleaning? • Date of the cleaning? • Locations cleaned? • Methods used to perform the cleaning? (40 CFR §§ 763.93(h) and 763.94(e)) Note: For details on initial cleaning after an inspection and before the initiation of any response action, other than O&M activities or repair, see 40 CFR § 763.91 (c)(1) and for details on any additional cleaning recommended by the management planner and approved by the LEA, see 40 CFR § 763.91 (c)(2). *Tip: See Model AMP Form 16 - Cleaning Record
☐ ☐ ☐	28. Does the management plan include a record of each O&M activity and major asbestos activity, with the following information: • Name of each person performing the activity? • For a major asbestos activity, the name, signature, state of accreditation and, if applicable, the accreditation number of each person performing the activity? • Start and completion date of each activity? • Location of the activity? • Description of the activity including preventative measures used? • If ACBM is removed, the name and location of the storage and disposal site for the ACM? (40 CFR §§ 763.93(h) and 763.94(f) and(g Note: The response actions for any maintenance activities disturbing friable ACBM, other than small-scale, short-duration maintenance activities, must be designed by persons accredited to design response actions and conducted by persons accredited to conduct response actions (40 CFR § 763.91(e)). Although not required, EPA suggests including in the AMP a copy of the accreditation. *Tip: See Model AMP Form 15 - Operations and Maintenance Activities
☐ ☐ ☐	29. Does the management plan include a record of each fiber release episode, whether major or minor, with the following information: • Date and location of the episode? • Method of repair? • Preventive measure or response action taken? • Name of each person performing the work? • If ACBM is removed, the name and location of the storage and disposal site of the ACM? (40 CFR §§ 763.93(h) and 763.94(h)) Note: A major fiber release episode is the falling or dislodging of more than 3 square or linear feet of friable ACBM (40 CFR § 763.91 (f)(2)). A minor fiber release episode is the falling or dislodging of 3 square or linear feet or less of friable ACBM (40 CFR § 763.91 (f)(1)). *Tip: See Model AMP Form 17 - Major/Minor Fiber Release Episode Log

Figure 6.14 *Cont'd*

Yes No N/A N/A - Not Applicable	School:

	Periodic Surveillance
☐ ☐ ☐	30. Does the management plan include a record of each periodic surveillance performed under 40 CFR § 763.92(b), with the following information: • Name of person performing the surveillance? • Date of the surveillance? • Any changes in the condition of the material? (40 CFR §§ 763.92(b)(2)(ii)-(iii), 763.93(h) and 763.94(d)) Note: A periodic surveillance of each school building must be conducted at least once every 6 months after a management plan has been in effect (40 CFR § 763.92(b)). *Tip: See Model AMP Form 18 - Periodic Surveillance Plan/Report

	Notification
☐ ☐ ☐	31. Does the management plan include the following notification information: • Description of the steps taken to notify, in writing, at least once a year, parent, teacher, and employee organizations of the availability of the management plan for review? • Dated copies of all such management plan availability notifications (e.g., letter, newsletter)? • Description of the steps taken to inform workers and building occupants, or their legal guardians, about inspections, reinspections, response actions, and post-response action activities, including periodic reinspection and surveillance activities that are planned or in progress? (Under 40 CFR § 763.84(c), the LEA must inform them about these activities at least once each school year.) (40 CFR §§ 763.93(e)(10) and 763.93(g)(4)) *Tip: See Model AMP Form 19 - Plan to Inform

Figure 6.14 *Cont'd*

Individual Assessment Form

AREA #:_____ AHERA CATEGORY #:_____

DESCRIPTION:_____

1. Location and Amount:_____

   _____

   _____

2. Condition, Type of Damage:_____

   _____

   _____

      Severity of Damage:_____

   _____

      Extent/Spread of Damage:_____

   _____

3. Accessibility: _____

   _____

   _____

4. Potential for Disturbance:_____

   _____

   _____

5. Causes of Damage:_____

   _____

   _____

6. Preventive Measures: _____

   _____

   _____

TYPE NAME:		SIGNATURE:	
ACCREDITATION AGENCY:	STATE:	ACCREDITATION:	DATE ISSUED:

Figure 6.15 Individual Assessment Form.

OKLAHOMA DEPARTMENT OF LABOR
O AND M, SSSD, and CLASS III PROGRAM SUBMITTAL

CONTRACTOR:	CONTRACTOR LICENSE #:
CONTRACTOR'S PROGRAM MANAGER:	CONTRACTOR PHONE #:
OWNER OF FACILITY:	OWNER'S DESIGNATED PERSON:
NAME OF FACILITY:	BUILDINGS TO BE INCLUDED: (Be specific as to buildings/addresses or areas to be included. Attach list if necessary)
FACILITY ADDRESS: FACILITY PHONE NUMBER:	

1. PROCEDURES FOR DESIGNATED PERSON:

    a.) If a fiber release occurs, the Designated Person will isolate the area and notify the Contractor, who will contact the Department of Labor when required.

    b.) The Designated Person will insure workers and building occupants are made aware of the presence and locations of asbestos containing materials. Occupants will be instructed not to disturb asbestos containing materials.

    c.) The Owner's Designated Person will insure that no unlicensed persons will be permitted to conduct any O & M activities.

2. CONTRACTOR'S RESPONSIBILITIES:

    a.) The Contractor will prepare all reports and collect the waste manifests and air monitoring reports to submit to the Department of Labor at the end of any month when work is conducted, within 30 days from the end of the month.

    b.) The Contractor will provide the Designated Person with a copy of the Monthly Reports to be maintained on site as required. The Designated Person will insert the reports into the Management Plan or the Asbestos Activity File and maintain these documents for a period of 30 years.

The Contractor or the Designated Person may contact the DOL when any unusual circumstances occur. The Contractor is ultimately responsible for making required notifications to the DOL.

This Contract is good for a period of _____ from the date of this Submittal to the Department of Labor.
(Specify "One Time" or "One Year")

Signatures:

_____   _____
Contractor's Program Manager   Date

_____   _____
Owner's Designated Person   Date

Revised: Jan 2011

Figure 6.16 Oklahoma Program Submittal form.

**Ten Leading Recommendations from *Asbestos Strategies* Process *(not ranked)*
Leading Short-Term Recommendations**

Action 1:	Update Existing Asbestos-in-Buildings Guidance		
Description:	Guidance documents provide workers with processes to follow in order to protect their health, protect the health of building occupants, and comply with regulations. EPA should update the "purple book" guidance document to make it the premier technical resource for managing asbestos in buildings and facilities, including industrial settings. The revised resource should include updated "green book" (operations and maintenance) information, and should be consistent with current federal regulations and good practices that have evolved since its release in 1985. The resulting resource, in a form such as an online integrated database of all relevant documents, will facilitate compliance with existing regulations, reducing asbestos exposure among contractors working in buildings.		
Lead Agency:	EPA	Supporting Agencies:	Occupational Safety and Health Administration (OSHA)
Action 2:	Encourage Compliance with Existing Regulations		
Description:	Regulatory agencies should encourage compliance with existing regulations and good practices for managing asbestos in buildings and conducting response actions. In some cases, businesses do not fully comply with existing regulations because they are not aware of the regulations. In other cases, they do not understand why it is important that the regulations be followed. Both may be addressed through a series of asbestos awareness seminars directed at the regulated community (building owners, contractors, and consultants). The seminars should be sponsored by EPA and OSHA, and hosted by the resident state asbestos authority. Joint sponsorship would be extremely valuable. Seminars should be held in conjunction with national or regional meetings of professional/trade associations, such as the Environmental Information Association (EIA), the International Facility Managers Association (IFMA), the Building Owners Management Association (BOMA), and the American Institute of Architects (AIA), to encourage participation by the target audience. Regulatory compliance will increase worker and building occupant safety, reduce asbestos exposure, and decrease costs associated with liability. This action should be undertaken in the context of a long-term effort to enforce existing regulations and improve consistency among agencies, as noted in Action 7 in Table 1.2.		
Lead Agency:	EPA	Supporting Agencies	OSHA, EIA, State Regulators
Action 3:	Clarify the Asbestos Definition to Address Asbestos Conamination in Vermiculite and Other Minerals		
Description:	Some of the asbestiform amphiboles found in the vermiculite from Libby, Montana, were not among the six minerals currently regulated as asbestos. Nevertheless, they were similar enough to regulated forms as to present dangerous health risks. The Libby vermiculite situation should be considered an important lesson, but not be treated as a typical case. A federal process should be undertaken promptly to clarify the definition of "asbestos." Many parties recommended that the definition should include all asbestiform amphiboles, in addition to currently regulated amphiboles and chrysotile. EPA, OSHA, and Mine Safety and Health Administration (MSHA) will need to evaluate how such a clarification should be accomplished and what consequences, if any, it would have on other industries. If adopted, this definition would enable federal agencies to address the risk of exposure from minerals such as winchite and richterite. USGS, trade associations, and other organizations can serve as resources for clarifying and understanding the science associated with creating a new definition.		
Lead Agency:	EPA	Supporting Agencies	MSHA, OSHA, USGS

Figure 6.17 Leading Short-Term Recommendations for Asbestos Strategies Process.

Action 4:	Advance a Federal Legislative Ban on Asbestos		
Description:	Asbestos continues to be used in products manufactured in the U.S. and in products imported into the U.S. This may present risk to workers or members of the public, and it increases the cost of regulatory compliance for building owners. A clearly defined legislative ban on the production, manufacture, distribution, and importation of products with commercially added asbestos is the most direct means to address concerns about remaining health risk and reduce future costs for facility owners and managers. Such a ban should be proposed by the Congress, promptly debated, and conclusively resolved. Enabling legislation would eliminate remaining products by a specified date, and installation of those products by a later date. Jurisdictional issues could be addressed in Congressional legislation that might not be achievable by individual agency rule-makings. Exceptions may be necessary for a small number of applications for which substitutes may not be available, and for research purposes. Implementing regulations, and perhaps the enabling legislation itself, could be challenged in the courts. A regulatory ban is within EPA's authority and is also an option. Many see a ban on asbestos, enacted to prevent future exposure, as a complementary action to a litigation resolution process that fairly compensates injuries resulting from past exposure.		
Lead Agency:	Congress	Supporting Agencies:	EPA, OSHA, U.S. Department of Commerce
Action 5:	Develop a National Mesothelioma Registry		
Description:	A national mesothelioma registry is necessary to facilitate epidemiological studies to evaluate the effects of asbestos exposure and enable public health officials to identify and respond to hazards. Many countries and some states have established mesothelioma registries. The establishment of such a registry would likely be performed by agencies within the Centers for Disease Control (CDC), including the National Center for Health Statistics, National Institute for Occupational Safety and Health, and the National Center for Environmental Health, in conjunction with Agency for Toxic Substances and Disease Registry (ATSDR) and state public health departments. An accompanying effort to connect interested parties with the best experts and data would improve research and treatment of asbestos-related disease.		
Lead Agency:	CDC	Supporting Agencies:	State Public Health Departments, ATSDR

Leading Long-Term Recommendations

Action 6:	Update Asbestos Model Training Curricula		
Description:	There have been substantial changes to federal regulations and standards since the model training curricula was developed. It is important to ensure that workers understand current regulations and understand why it is important to follow these regulations. EPA should update the model training curricula to ensure that all relevant agencies' priorities are reflected. Updating the training will make the curricula consistent with existing regulations and increase worker safety. The updated versions should cover the revised OSHA asbestos standards, revised EPA asbestos National Emission Standard for Hazardous Air Pollutants (NESHAP) standards, the EPA Worker Protection Rule, new respirator designations/regulations, and other topics. The training providers should also be permitted to vary the course content in refresher courses.		
Lead Agency:	EPA	Supporting Agencies:	State Regulators, OSHA
Action 7:	Enforce Existing Asbestos Regulations		
Description:	Existing asbestos regulations have been designed to reduce the adverse effects from asbestos exposure on the health of the public and of workers. Inconsistent interpretation leads to confusion; lax enforcement allows substandard practices. Both can lead to increased health risk as regulations are ignored. EPA, OSHA, Consumer Product Safety Commission (CPSC), and state regulators should focus on more stringent, predictable, and consistent enforcement of these existing regulations, which may offer greater benefit than committing scarce resources to new rule-making efforts. This recommendation can be implemented immediately; however, such an effort must continue for the long term. Consistent interpretations and streamlining across agencies will lead to increased compliance and potential reduced liability for businesses. Any step that EPA and OSHA can take to encourage the enforcement of existing regulations at the local level will likely prove most effective. To this end, consideration should be given to the use of a form such as the one created by EIA to assure compliance with existing regulations at the time applications are made for building, renovation, or demolition permits. This action ties into Action 2 in Table 1.1.		
Lead Agency:	EPA	Supporting Agencies:	OSHA, CPSC State Regulators

Figure 6.17 *Cont'd*

Action 8:	Reduce the Occurrence of Unintended Asbestos in Products		
Description:	Accidental contamination of mineral products with asbestos can increase risks to the users of these products or the workers who process them, and in turn can result in major liability losses affecting the mineral product companies. Assisting companies in avoiding asbestos in the first places is in the best interest of all parties. Reduction of naturally occurring asbestos in products could be achieved by a program set up by a consortium of mining concerns to develop a sampling and analytical protocol to analyze bulk materials at the mining stage for chrysotile and all asbestiform amphibole forms of asbestos. Oversight of such a program may be provided by EPA and MSHA, with technical assistance by the National Institute for Occupational Safety and Health (NIOSH), the National Institute for Standards and Technology (NIST), and USGS. This program would assist the mining and quarrying industries in avoiding unwanted asbestos in their products. The program would provide a degree of assurance to users of these raw materials that they are not contaminated with asbestos.		
Lead Agency:	EPA	Supporting Agencies:	Mining Industry, MSHA, NIOSH, NIST, USGS
Action 9:	Address Asbestos-Containing Products in Commerce		
Description:	Consumers, employers, and building owners are in many cases unaware of the inclusion of asbestos in products. Without this knowledge, they cannot take appropriate steps to protect their health. A coordinated effort to educate consumers, employers, and building owners about products with commercially-added asbestos is necessary. Such a program would assist the target audience in making an informed decision about which products are legally available with commercially added asbestos. This education and outreach effort would be performed by EPA, OSHA, and CPSC. These agencies would need to perform research to determine which products actually have commercially added asbestos, which do not, and which are to be phased out voluntarily by manufacturers. Congress should consider amending the Asbestos Information Act of 1988 to require manufacturers and importers to update information on their asbestos-containing products to EPA.		
Lead Agency:	EPA	Supporting Agencies:	CPSC, OSHA, Congress, Bureau of Customs and Border Protection
Action 10:	Partner with State Agencies in Support of Asbestos Training		
Description:	Training requirements for contractors must be enforced. Training fraud does exist and is a real concern, particularly with some contractors producing fraudulent certification. If untrained contractors perform asbestos abatement, they put themselves and building occupants at risk. Training providers under the EPA model accreditation plan (MAP) and corresponding state plans should be audited with sufficient frequency to ensure that training is provided, tests are conducted, records are maintained, and certificates are issued. This action, conducted in concert with Action 6, will increase worker safety and the effectiveness of abatement efforts. Reducing the incidence of training fraud will provide greater security to building occupants and owners. Partnering with state agencies will provide better coordination.		
Lead Agency:	EPA	Supporting Agencies:	State Regulators, Training Providers, OSHA

Figure 6.17 *Cont'd*

Summary of When to Apply Key O&M Work Practices

	Likelihood of ACM Disturbance		
	Contact Unlikely	Accidental Disturbance Possible	Disturbance Intended or Likely
Management Responsibilities			
Need Pre-Work Approval from Asbestos Program Manager	Review by Program Manager	Yes	Yes
Special Scheduling or Access Control	No	Yes	Yes
Supervision Needed	No	Initial, At Least	Yes
HVAC System Modification	None	As Needed	Shut Down
Area Containment	None	Drop cloths, Mini-enclosures	Yes
Personal Protection			
Respiratory Protection	Available for Use	Yes	Yes
Protective Clothing	None	Review by Asbestos Program Manager	Yes
Work Practices			
Use of Wet Methods	No	As Needed	Yes
Use of HEPA Vacuum	Available for Use	Available for Use	As Needed
1) In the area where work takes place 2) Type of containment may vary. For example, small-scale, short-duration tasks may not require full containment.			

Figure 6.18 The EPA's Summary of When to Apply Key O&M Work Practices.

equipment. Like any other subcontractors you are accustomed to managing, you can use the same fundamentals when supervising asbestos workers. Most important, do not become more involved in an asbestos situation than what you are qualified to do. You will find that some of the sample illustrations in Figures 6.21 through 6.24 could help you decide whether getting certified is right for you.

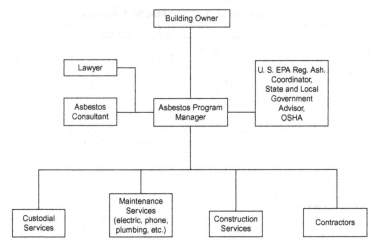

Figure 6.19 Sample management organization plan.

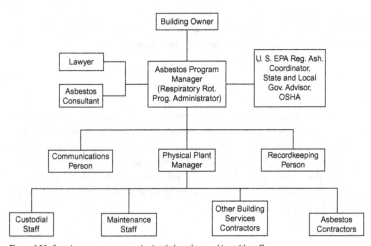

Figure 6.20 Sample management organizational plan when working with staff.

Box 6.4. Key Points about LEA Responsibilities

The LEA must have an accredited inspector conduct **inspections** of each school building under its authority. A **reinspection** of all friable and nonfriable known or assumed ACBM in each school building must be conducted at least once every 3 years that a management plan is in effect. A management planner must review all 3-year inspection reports.

For each inspection and reinspection, an accredited inspector must provide a written **assessment** of all friable known or assumed ACBM in the school building.

The LEA must have an accredited management planner review the results of the inspection/reinspection and the assessment and make written recommendations on appropriate response actions. The accredited management planner also prepares the asbestos **management plan** for each school under its authority.

The LEA must select the appropriate **response actions** consistent with the assessment of the ACBM and the recommendations of the management planner.

The LEA must implement an **operations and maintenance (O&M) program** whenever any friable ACBM is present or assumed to be present in a building under its authority.

Building inspectors, management planners, project designers, contractors/ supervisors, and asbestos workers must complete EPA- or state-approved courses and receive accreditation before they can perform any asbestos-related activities. The AHERA Rule also specifies training requirements for LEA designated persons and custodial and maintenance workers, although these individuals are not required to complete any EPA-approved courses or receive accreditation.

The LEA must conduct **periodic surveillance** in each building under its authority at least once every 6 months after a management plan is in effect.

The LEA must comply with the requirements to provide **notification** about asbestos activities to workers, students, parents, teachers, and short-term workers.

The LEA must maintain **records** in accordance with the AHERA regulations.

The LEA must attach a **warning label** immediately adjacent to any friable and nonfriable ACBM and assumed ACBM located in routine maintenance areas (such as boiler rooms) at each school building.

Box 6.5. Checklist of the Local Education Agency's General Responsibilities under AHERA

The AHERA Designated Person must complete and sign a statement that the Local Education Agency has met (or will meet) the following responsibilities. All references are to specific provisions to the AHERA regulations (under § 763.84). The AHERA Designated Person should be able to answer "yes" to each statement that follows.

1. The activities of any persons who perform inspections, reinspections, and periodic surveillance; develop and update management plans; and develop and implement response actions, including operations and maintenance, are carried out in accordance with 40 CFR, Part 763, Subpart E.

2. All custodial and maintenance employees are properly trained as required in 40 CFR, Part 763, Subpart E, and all other applicable federal and/or state regulations (e.g., the Occupational Safety and Health Administration Asbestos Standard for Construction, the EPA Worker Protection Rule, or applicable state regulations).

3. All workers and building occupants, or their legal guardians, are informed at least once each school year about inspections, response actions, and postresponse action activities, including periodic reinspections and surveillance activities, that are planned or in progress.

4. All short-term workers (e.g., telephone repair workers, utility workers, or exterminators) who may come in contact with asbestos in school are provided information regarding the locations of ACBM and assumed ACBM.

5. All warning labels are posted in accordance with § 763.95.

6. All management plans are available for inspection, and notification of this availability has been provided in accordance with § 763.93(g).

7. The undersigned person designated by the LEA according to § 763.84(g)(1) has received adequate training as required by § 763.84(g)(2).

8. The LEA has and will consider whether any conflict of interest may arise from the interrelationship between accredited personnel, and whether this potential conflict should influence the selection of accredited personnel to perform activities under 40 CFR, Part 763, Subpart E.

Box 6.6. Implementation Requirements for Operations Associated with the Management Plan

REQUESTS	DEADLINE
The Management Plan	The plan must be kept current with ongoing O&M, periodic surveillance, inspection, reinspection, and response action activities, including updating the locations of ACBM after response actions and O&M activities.
O&M Program	Must begin immediately upon the identification of any friable ACBM present or assumed to be present in the building.
O&M Training	To work in a building that may contain asbestos, custodial workers and maintenance staff members must have completed the 2-hour training class described in § 763.92(a)(1) within 60 days of employment. Workers must have completed the 14-hour training requirement described in § 763.92(a)(2) to conduct O&M activities that may disturb ACBM.
Periodic Surveillance	Under § 763.92(b)(1) of the AHERA Rule, periodic surveillance must be conducted at least once every 6 months after a management plan is in effect.
Warning Labels	Must be posted as soon as possible after identification of ACBM in any routine maintenance area.
Management Plan Availability for Public Review	The plan must have been made available for public review in the administrative office of the LEA on the date on which it was submitted to the governor for review. Notification of the plan availability must be made annually.
Isolate a Functional Space with Significantly Damaged Friable Surfacing ACBM	Must be isolated immediately and access restricted if such measures are needed to protect human health and the environment.
Repair and Maintain Damaged or Significantly Damaged TSI	Must begin as soon as a management planner and LEA determine that these conditions exist.

Box 6.7. Frequent Problems with Management Plans

The Asbestos Management Plans (Plan) should be considered a "living" document. *Some Plans are left exactly the same as they were when they were created, with no updates whatsoever.* This is particularly true with respect to required records of periodic surveillances, annual notifications, response actions or fiber releases, and for records of the 2-hour and 16-hour training for school employees and maintenance workers. In fact, the *administrative staff at individual schools are sometimes unaware of the existence of management plans and/or do not know where the school's copy of the plan is kept.*

Copies of all pertinent certification credentials for AHERA inspectors, management planners, project designers, workers, and supervisors who have participated in any response actions are required to be in the management plan, but *are not always included.* Also *proper documentation of air samplers' and laboratories' accreditations are sometimes missing* from Plans.

Homogeneous areas are often not clearly (and frequently are not properly) defined on the basis of color, texture, and size. Plaster and sheetrock are probably the most often overlooked materials that are likely to comprise major areas of suspected asbestos-containing building materials (ACBM).

Sampling locations within the individual homogeneous areas are often not described precisely enough to provide for any relocation of individual original sampling sites with any degree of certainty.

Frequently *insufficient numbers of samples are collected from individual homogeneous areas* (the correct minimum number being dependent on the type of building material and the homogeneous area size), and the sites for the sampling that was done may have been selected in a manner other than as is set forth in the management plan for how sampling locations were to have been determined. Also, where *warning signs* are required, they may be *missing*, or if present, they may not employ the prescribed text.

Sometimes *functional areas are not taken into consideration in the preparation of assessment and response actions recommendations.* Also recommended response actions may not have been carried out according to schedules shown in the management plans and explanations or changes in the schedules may be absent.

Portable buildings on school grounds are sometimes overlooked in management plans, or these units may have been moved onto or off of a school's grounds without the school's management plan having been updated.

Box 6.8. Key Points about the Management Plan

The management plan is a **site-specific guidance document** that the LEA designated person must follow in managing the ACBM present in a school building.

The management plan must be prepared by an **accredited management planner** and must be updated in a timely manner.

The management plan must include the documentation required under § 763.87 of the AHERA Rule for each laboratory performing a bulk sample analysis and the results of each analysis.

In the management plan, the management planner must recommend an **appropriate response action** (operations and maintenance, repair, encapsulation, enclosure, or removal) for all areas of TSI and friable ACBM (including ACBM that has the potential of becoming friable).

All of the initial response actions implemented to control friable asbestos require a **project design** specifying how to conduct the abatement project.

Final air clearance of a functional space after a response action to remove, encapsulate, or enclose ACBM involves a **visual inspection** and the collection and analysis of **air samples.**

Final air sampling must be done using the transmission electron microscopy (TEM) method, unless the project involves no more than 160 square feet or 260 linear feet, in which case phase contrast microscopy (PCM) may be used.

The LEA designated person is responsible for ensuring that the activities related to the management plan are implemented and that the management plan is updated in a timely manner.

Box 6.9. Key Points about Reinspections and Periodic Surveillance

As long as any ACBM remains in a school building, the building must be **reinspected** at least once every 3 years.

The reinspection and assessments/reassessments must be conducted by an **accredited inspector.** The results of the inspection must be submitted to the Designated Person within 30 days to include into the management plan.

The **management planner** must: (1) review the results of the reinspection, (2) make written response action and preventive measure recommendations for each area of friable surfacing and miscellaneous ACBM and each area of TSI ACBM, (3) determine whether additional cleaning is necessary and, if so, specify how, when, and where to perform cleaning, (4) include an implementation schedule for the recommended activities and make an estimate regarding the resources needed to conduct the activities, and (5) review the adequacy of the Operations & Maintenance Program.

At least once every 6 months after a management plan is in effect, the LEA must conduct **periodic surveillance** in each building that contains ACBM or is assumed to contain ACBM.

Oklahoma Department of Labor

Mark Costello
COMMISSIONER OF LABOR

ASBESTOS INSPECTOR APPLICATION

☐ NEW   ☐ RENEWAL   LICENSE #

1. Applicant's Name				
2. Home Address		3. City	4. State	5. Zip
6. Date of Birth	7. Social Security Number	8. Phone		
8. Hair Color	9. Eye Color	10. Weight	11. Height	
12. Company Name				
2. Company Address		3. City	4. State	5. Zip
17. Company Phone		18. Company Contact Person		

Do you currently hold any other Oklahoma Asbestos Licenses? ☐ YES   ☐ NO
If yes, please indicate the type of license and the license number:

CHECK TYPE LICENSE	LICENSE NUMBER		CHECK TYPE LICENSE	LICENSE NUMBER
Supervisor			Worker	
Mgmt. Planner			Project Designer	

Please check one of the following:

	NON-EXEMPT	Submit check or money order for $25.00 made payable to the Oklahoma Department of Labor.

OR

	EXEMPT	If the applicant is the employee of a political subdivision of the State of Oklahoma and agrees to perform any AHERA asbestos work on behalf of that employer only, no fee is required.

FOR OFFICE USE ONLY				
Date	DEO	Receipt No.	License No.	Asbestos Admin Approval/Date
Type of Payment	1 2 3	#	Endorser	

Revised 2011-Jan   3017 N. STILES OKLAHOMA CITY, OKLAHOMA 73105 • www.labor.ok.gov • PHONE: (405)521-6464 • FAX (405) 521-6016
Toll Free Number (888) 269-5353   Page 1

Figure 6.21 Oklahoma's Asbestos Inspector Application.

ALL APPLICANTS MUST SUBMIT THE FOLLOWING:
1. A copy of your current drivers license or state issued photo identification card.
2. A copy of your social security card or letter from social security administration showing you have rapplied for your card.
3. Copy(s) of current refresher training class*

NEW APPLICANTS MUST ALSO SUBMIT THE FOLLOWING, IN PERSON:
1. Copies of Original AHERA Inspector training class and all subsequent refresher courses.*
2. Provide current drivers license or state issued photo identification card.
3. Provide social security card or letter from social security administration showing you have reapplied for your card.
4. Affidavit regarding citizenship.

*Training must have been provided by an educational institution, government agency or labor union and must have been accredited by the U.S. Environmental Protection Agency.

I hereby authorize the educational institutions to release verification of completion of the courses presented in this application.

I further affirm, upon my oath, to follow Title 40 of the Oklahoma Statutes, Section 450 through 456 as amended, and the Abatement of Friable Asbestos Materials Rules OAC 380, Chapter 50. I understand that a violation of any law or rule may subject my license to be suspended or revoked, or may subject me to cease and desist orders, injunctive measures, and criminal penalties for criminal violations.

I, upon my oath, do state that the above information is a true statement, and further state that I am not under any type of disciplinary action, including license revocation or suspension, by any State or political division thereof, or by the United States government, for any illegal or improper activity, civil or criminal, involving asbesto-containing material.

_____   _____
Applicants Signature   Date

THIS APPLICATION MUST BE NOTARIZED
Acknowledged

County_____
State _____

Signed or affirmed before me this _____ day of _____, _____.

My Commission Expires _____

_____
Notary Public

OAC 380:50-5-10. Licensing of Asbestos Abatement Inspectors: Licensing requirements for AHERA Asbestos Inspectors are as follows: (1) Inspection for asbestos-containing materials in any facility under the jurisdiction of OAC, Sections 450 through 456 shall be performed only by persons who are licensed as AHERA Inspectors by the Oklahoma Department of Labor. (2) AHERA Inspectors shall be licensed as a special category of Asbestos Worker and shall have completed a 24-class-hour course for AHERA Inspectors all subsequent Asbestos Inspector refresher training which fully meet the requirement of OAC 380:50-6-4 and 380:50-6-9. (3) Applications shall be submitted on forms prescribed by the Commissioner. (4) The license fee shall be twenty-five dollars ($25.00) per year. (5) The license shall be issued in the name of the individual applicant. (6) The license shall be issued for a period not to exceed one year and shall expire concurrently with the asbestos training and subsequent refresher training. There will be no grace period wherein a Inspector will be allowed to work with an expired license. (7) Any Inspector who has not taken the required AHERA Inspector refresher training course within two years of the previos Inspector training or refresher course, shall repeat the AHERA Inspector training requirement of OAC 380:50-6-4 and 380:50-6-9. (8) The license shall be issued in the name of the individual applicant. (9) License cards certificates shall be available at the job site for inspection by the Department of Labor.

OAC 380:50-6-4. Initial training for Asbestos Inspectors: Training requirements are as follows: (1) Training shall be supplied by an EPA or DOL accredited educational institution, labor union, or government agency, or from a private vocational education provider licensed by the state where it operates (pursuant to 70O.S. § 21-103 within the state of Oklahoma and accredited by EPA or an EPA approved govermental agency. (2) Training providers must have EPA approval or approval by a state that has an approved Model Accreditation Plan (MAP) that is as stringent or exceeds the minimum requirements of the EPAMAP. (3) AHERA Inspector's course shall be no less than three days in length and shall include: lectures, demonstrations, and at least 4 hours of hands-on training, individual respirator fit testing, course review and an written examination. (4) Asbestos Abatement Inspector training courses shall consist of the training required in OAC 380:50-6-4 and 380:50-6-9.

Figure 6.21 *Cont'd*

Oklahoma Department of Labor

Mark Costello
COMMISSIONER OF LABOR

ASBESTOS
PROJECT DESIGNER APPLICATION

☐ NEW   ☐ RENEWAL   LICENSE #

1. Applicant's Name			

2. Home Address			

3. City	4. State	5. Zip	

6. Date of Birth	7. Social Security Number	8. Phone	

8. Hair Color	9. Eye Color	10. Weight	11. Height

12. Doing businees as:			

13. At the address of:			

14. City	15. State	16. Zip	

17. Company Phone	18. Company Contact Person	

FOR OFFICE USE ONLY					
Date	DEO	Receipt No.	License No.		Asbestos Admin Approval/Date
Type of Payment	1 2 3	#		Endorser	

Figure 6.22 Oklahoma's Asbestos Project Designer Application.

Do you currently hold any other Oklahoma Asbestos Licenses?

☐ YES   ☐ NO

If yes, please indicate the type of license and the license number:

CHECK TYPE LICENSE	LICENSE NUMBER
Supervisor	
Mgmt. Planner	

CHECK TYPE LICENSE	LICENSE NUMBER
Worker	
Inspector	

ALL APPLICANTS MUST SUBMIT THE FOLLOWING:

1. A copy of your current drivers license or state issued photo identification card.
2. A copy of your social security card or letter from social security administration showing you have reapplied for your card.
3. Copy(s) of current refresher training class*

NEW APPLICANTS MUST ALSO SUBMIT THE FOLLOWING, IN PERSON:

1. Copies of Original AHERA Project Designer training class and all subsequent refresher courses.*
2. Provide current drivers license or state issued photo identification card.
3. Provide social security card or letter from social security administration showing you have reapplied for your card.
4. Affidavit regarding citizenship.

NEW APPLICANTS MUST ALSO SUBMIT THE FOLLOWING:

1. Copies of Official Transcript of diploma/degree in architecture, engineering or industrial hygiene (or an equivalent)

    OR

2. Degree of Equivalence application (see above NOTE)

NOTE: If applicant does not have the required degree/diploma, please fill out a Degree of Equivalence application form and submit it with this application for approval.

*Training must have been provided by an educational institution, government agency or labor union and must have been accredited by the U.S. Environmental Protection Agency.

Education Information

List schools attended and applicable training received. Provide additional information on a separate sheet if necessary.

High School	Name_____	Year of Graduation_____
(submit copy of GED if appropriate)	City _____	State_____

College or University	Location (City/State)	From: Mo/Yr	To: Mo/Yr	Diploma or Degree	Major Subjects	Total Credit Hrs

Figure 6.22 *Cont'd*

If applicant is the employee of a political subdivision of the State of Oklahoma and agrees to perform any AHERA Inspection or Management Planning services on behalf of that employer only, applicant is **EXEMPT** and no fee is required. Otherwise, applicant is considered **NON-EXEMPT** and a fee of $500.00 is required.

Select One: □ **EXEMPT**   □ **NON-EXEMPT**

I hereby authorize the above named educational institutions to release verification of completion of the AHERA Inspector/Management Planner courses presented in this license application.

I further affirm, upon my oath, to follow Title 40 of the Oklahoma Statutes, Section 450 through 456 as amended, and the Abatement of Friable Asbestos Materials Rules OAC 380, Chapter 50. I understand that a violation of any law or rule may subject my license to be suspended or revoked, or may subject me to cease and desist orders, injunctive measures, and criminal penalties for criminal violations.

I, upon my oath, do state that the above information is a true statement, and further state that I am not under any type of disciplinary action, including license revocation or suspension, by any State or political division thereof, or by the United States government, for any illegal or improper activity; civil or criminal, involving asbestos-containing material.

_____   _____
Applicant's Signature   Date

THIS APPLICATION MUST BE NOTARIZED

Acknowledged

County _____

State _____

   Signed or affirmed before me this _____ day of _____ , _____ .

                                                          _____
                                                          Notary Public

My Commission Expires _____

Figure 6.22 *Cont'd*

OAC 380:50-5-12. of the <u>Abatement of Friable Asbestos Materials Rules</u> states: Licensing of
AHERA Project Designers.

Licensing requirements for AHERA Project Designers are as follows: (1) Preparation of plans and/or
specifications for response actions for asbestos-containing materials in any facility under the jurisdiction
of Title 40, Sections 450 through 456 shall be performed only by persons who are licensed as AHERA
Project Designers by the Oklahoma Department of Labor. (2) AHERA Project Designers shall have met
all requirements for accreditation for Asbestos Abatement Contractor or Project Designer, and in
addition, shall have a bachelor's or advanced degree in architecture, engineering, or industrial hygiene,
or an equivalent combination of education, training, and experience as determined by the Commissioner.
(3) Applications shall be submitted on forms prescribed by the Commissioner. (4) The license fee shall
be Five Hundred Dollars ($500.00) per year. If the applicant holds a current AHERA Management
Planner license, there shall be no additional fee charged. (5) The license shall be issued in the name
of the individual applicant. (6) The license shall be issued for a period not to exceed one year and shall
expire concurrently with the initial Asbestos Contractor of Project Designer training and subsequent
Project Designer refresher training. There will be no grace period wherein a Project Designer will be
allowed to work with an expired license. (7) Any Project Designer who has not taken the required
AHERA Project Designer refresher training course within two years of the previous initial Contractor or
Project Designer training or Project Designer refresher course shall repeat the initial Project Designer
training requirements of **OAC 380:50-6-6 and 380:50-6-11.**

OAC 380:50-6-6 of the <u>Abatement of Friable Asbestos Materials</u> states:
Training Requirements:
Training requirements are as follows: (1) Training shall be supplied by an EPA or DOL accredited,
educational institution, labor union, or government agency, or from a private vocational education
provider licensed by the state where it operates (pursuant to 70 O.S.§ 21-103 within the state of
Oklahoma) and approved by EPA or an EPA-approved governmental agency. (2) Training providers
must have EPA approval or approval by a State that has an approved Model Accreditation Plan (MAP)
that is as stringent or exceeds the minimum requirements of the EPA MAP. (3) The AHERA Project
Designer's course shall be no less than three days in length and shall include lectures, demonstrations,
a field trip, course review and a written examination. (4) Asbestos Abatement Project Designer training
courses shall consist of the training required in **OAC 380:50-6-6 and 380:50-6-11.** (5) In addition to the
training required for an AHERA Project Designer, persons seeking accreditation or licensure must also
have a minimum of a bachelor's or advanced degree in architecture, engineering or industrial hygiene
or an equivalent combination of education, training and experience as determined by the Commissioner
of Labor.

Figure 6.22 *Cont'd*

Oklahoma Department of Labor

Mark Costello
COMMISSIONER OF LABOUR

ASBESTOS
MANAGEMENT PLANNER APPLICATION

☐ NEW   ☐ RENEWAL   LICENSE #

1. Applicant's Name			
2. Home Address			
3. City	4. State	5. Zip	
6. Date of Birth	7. Social Security Number	8. Phone	
8. Hair Color	9. Eye Color	10. Weight	11. Height
12. Doing Business as:			
13. At the Address of:			
14. City	15. State	16. Zip	
17. Company Phone		18. Company Contact Person	

3017 N. STILES, OKLAHOMA CITY, OKLAHOMA 73105 • www.laobr.ok.gov • PHONE: (405)521-6464 • FAX (405) 521-6016
Revised 2011-January   Toll Free Number (888) 269-5353   Page 1

Figure 6.23 Asbestos Management Planner Application.

Do you currently hold any other Oklahoma Asbestos Licenses? ☐ YES   ☐ NO

If yes, please indicate the type of license and the license number:

CHECK TYPE LICENSE	LICENSE NUMBER		CHECK TYPE LICENSE	LICENSE NUMBER
Supervisor			Inspector	
Mgmt. Planner			Project Designer	

ALL APPLICANTS MUST SUBMIT THE FOLLOWING:

1. A copy of your current drivers license or state issued photo identification card.
2. A copy of your social security card or letter from social security administration showing you have reapplied for your card.
3. Copy(s) of current refresher training class*

NEW APPLICANTS MUST ALSO SUBMIT THE FOLLOWING, IN PERSON:

1. Copies of Original AHERA Inspector training class and all subsequent refresher courses.*
2. Provide current drivers license or state issued photo identification card.
3. Provide social security card or letter from social security administration showing you have reapplied for your card.
4. Affidavit regarding citizenship.

NEW APPLICANTS MUST ALSO SUBMIT THE FOLLOWING:

1. Copies of Original Inspector/Management Planner training course and all subsequent refresher courses.*
2. Copies of Official Transcript of technical diploma/degree (see NOTE below)
    OR
    Degree of Equivalence application (see NOTE below)

NOTE: If applicant does not have a technical degree/diploma, please fill out a Degree of Equivalence application form and submit it with this application for approval.

*Training must have been provided by an educational institution, government agency or labor union and must have been accredited by the U.S. Environmental Protection Agency

---

Education Information

List schools attended and applicable training received. Provide additional information on a seperate sheet if necessary.

High School   Name _____ Year of Graduation _____
(submit copy of GED if appropriate) City _____ State _____

College or University	Location (City/State)	From: Mo/Yr	To: Mo/Yr	Diploma or Degree	Major Subjects	Total Credit Hrs

Figure 6.23 *Cont'd*

If applicant is the employee of a political subdivision of the State of Oklahoma and agrees to perform any AHERA Inspection or Management Planning services on behalf of that employer only, applicant is **EXEMPT** and no fee is required. Otherwise, applicant is considered **NON-EXEMPT** and a fee of $500.00 is required.

Select One: □ **EXEMPT**   □ **NON-EXEMPT**

I hereby authorize the above named educational institutions to release verification of completion of the AHERA Inspector/Management Planner courses presented in this license application.

I further affirm, upon my oath, to follow Title 40 of the Oklahoma Statutes, Section 450 through 456 as amended, and the <u>Abatement of Friable Asbestos Materials Rules OAC 380, Chapter 50</u>. I understand that a violation of any law or rule may subject my license to be suspended or revoked, or may subject me to cease and desist orders, injunctive measures, and criminal penalties for criminal violations.

I, upon my oath, do state that the above information is a true statement, and further state that I am not under any type of disciplinary action, including license revocation or suspension, by any State or political division thereof, or by the United States government, for any illegal or improper activity, civil or criminal, involving asbestos- containing material.

_____   _____
Applicant's Signature   Date

THIS APPLICATION MUST BE NOTARIZED

<u>**Acknowledged**</u>

County _____

State _____

  Signed or affirmed before me this _____ day of _____, _____ .

_____
Notary Public

My Commission Expires _____

Figure 6.23 *Cont'd*

OAC 380:50-5-11. Licensing of Asbestos Abatement Management Planner:
(1) Preparation of management plans specifying response actions for asbestos-containing materials in any facility under the jurisdiction of Title 40, Sections 450 through 456 shall be performed only by persons who are licensed as AHERA Management Planners by the Oklahoma Department of Labor. (2) AHERA Management Planners shall be licensed as a special category of Asbestos Contractor, shall have a bachelor's degree in a technical subject, or equivalent, and, in addition to the AHERA Inspector training outlined in **OAC 380:50-6-4 and 380:50-6-9,** shall have completed a 16-hour course for AHERA Asbestos Management Planners that fully meets the requirements of **OAC 380:50-6-5 and 380:50-6-10.** (3) Applications shall be submitted on forms prescribed by the Commissioner. (4) The license fee shall be Five Hundred Dollars ($500.00) per year. If the applicant holds a current AHERA Project Designer license, there shall be no additional fee charged. (5) The license shall be issued in the name of the individual applicant. (6) The license shall be issued for a period not to exceed one year and shall expire concurrently with the initial Management Planner training and subsequent Management Planner refresher training. (7) Any Management Planner who has not taken the required AHERA Management Planner refresher training course within two years of the previous Management Planner training or refresher course shall repeat the AHERA Inspector training requirements of **OAC 380:50-6-4 and 380:50-6-9.**

OAC 380:50-6-5. Initial training for Asbestos Management Planners:
Training requirements are as follows: (1) Training shall be supplied by an EPA- or DOL-accredited educational institution, labor union, or government agency, or from a private vocational education provider licensed by the state where it operates (pursuant to 70 O.S. § 21-103 within the state of Oklahoma and accredited by EPA or an EPA-approved governmental agency. (2) Training providers must have EPA approval or approval by a state that has an approved Model Accreditation Plan (MAP) that is as stringent or exceeds the minimum requirements of the EPA MAP. (3) All persons seeking accreditation, as a Management Planner shall complete a three-day Inspector training course and accreditation, as a prerequisite to the two-day Management Planners course. (4) Asbestos Abatement Inspector training courses shall consist of the training required in **OAC 380:50-6-4 and 380:50-6-9.** (5) Asbestos abatement Management Planner training courses shall consist of the training required in **OAC 380:50-6-5 and OAC 380:50-6-10.** (6) In addition to the training required for an AHERA Management Planner, persons seeking licensure in the State of Oklahoma shall also have a minimum of a bachelor's degree in engineering, industrial hygiene or other advanced fields, or an equivalent combination of experience, education and training as determined by the Commissioner of Labor.

Figure 6.23 *Cont'd*

Oklahoma Department of Labor

Mark Costello
COMMISSIONER OF LABOR

ASBESTOS ABATEMENT
SUPERVISOR APPLICATION

☐ NEW ☐ RENEWAL LICENSE #

1. Applicant's Name			
2. Home Address			
3. City	4. State	5. Zip	
6. Date of Birth	7. Social Security Number		8. Phone
8. Hair Color	9. Eye Color	10. Weight	11. Height
12. Company Name			
13. Company Address			
14. City	15. State	16. Zip	
17. Company Phone		18. Company Contact Person	

ALL APPLICANTS MUST SUBMIT THE FOLLOWING:

1. A copy of your current drivers license or state issued photo identification card.
2. Copy(s) of current refresher training class*
3. Copy(s) of Cardiopulmonary Resuscitation (CPR) training class
4. Copy(s) of First Aid class
5. A copy of your social security card or letter from social security administration showing you have reapplied for your card.

NEW APPLICANTS MUST ALSO SUBMIT THE FOLLOWING, IN PERSON:

1. Copies of Original Worker or Contractor/Supervisor training class and all subsequent refresher courses*
2. Copies of Confined Space Entry*
3. Copies of NIOSH 582 or Air Monitoring*
4. Affidavit regarding citizenship.

*Training must have been provided by an educational institution, government agency or labor union and must have been accredited by the U.S. Environmental Protection Agency.

FOR OFFICE USE ONLY				
Date	DEO	Receipt No.	License No.	Asbestos Admin Approval/Date
Type of Payment	1 2 3	#	Endorser	

3017 N.STILES, OKLAHOMA CITY, OKLAHOMA 73105 • www.labor.ok.gov • PHONE: (405)521-6464 • FAX (405) 521-6016
Toll Free Number (888) 269-5353

Revised 2011-Jan

Page 1

Figure 6.24 Asbestos Abatement Supervisor Application.

Do you currently hold any other Oklahoma Asbestos Licenses?   ☐ YES   ☐ NO

If yes, please indicate the type of license and the license number:

CHECK TYPE LICENSE	LICENSE NUMBER
Worker	
Mgmt. Planner	

CHECK TYPE LICENSE	LICENSE NUMBER
Inspector	
Project Designer	

Please check one of the following:

	NON-EXEMPT	Submit check or money order for $25.00 made payable to the Oklahoma Department of Labor
	OR	
	EXEMPT	If the applicant is the employee of a political subdivision of the State of Oklahoma and agrees to perform any AHERA asbestos work on behalf of that employer only, no fee is required

Do you have any asbestos violations?   ☐ YES   ☐ NO

If yes, please list job and type of violation: _____

_____

_____

Abatement Projects
ONLY NEW APPLICANTS MUST COMPLETE THIS SECTION
List six (6) abatement projects on which you worked:

1. Employer: _____
   Phone: _____
   Project name: _____
   Dates from/to: _____
   Contact Person: _____

2. Employer: _____
   Phone: _____
   Project name: _____
   Dates from/to: _____
   Contact Person: _____

3. Employer: _____
   Phone: _____
   Project name: _____
   Dates from/to: _____
   Contact Person: _____

4. Employer: _____
   Phone: _____
   Project name: _____
   Dates from/to: _____
   Contact Person: _____

5. Employer: _____
   Phone: _____
   Project name: _____
   Dates from/to: _____
   Contact Person: _____

6. Employer: _____
   Phone: _____
   Project name: _____
   Dates from/to: _____
   Contact Person: _____

Employment History
ONLY NEW APPLICANTS MUST COMPLETE THIS SECTION
If you have more than three separate previous employments, please sign and attach additional sheets in the same format as below:

Figure 6.24 *Cont'd*

| Employer's Name: |
| Employer's Address: |
| Project Name: |
| Dates from/to: |
| Contact Person: |
| Employer's Name: |
| Employer's Address: |
| Project Name: |
| Dates from/to: |
| Contact Person: |
| Employer's Name: |
| Employer's Address: |
| Project Name: |
| Dates from/to: |
| Contact Person: |

I hereby authorize the educational institutions to release verification of completion of the courses presented in this application.

I further affirm, upon my oath, to follow Title 40 of the Oklahoma Statutes, Section 450 through 456 as amended, and the Abatement of Friable Asbestos Materials Rules OAC 380, Chapter 50. I understand that a violation of any law or rule may subject my license to be suspended or revoked, or may subject me to cease and desist orders, injunctive measures, and criminal penalties for criminal violations.

I, upon my oath, do state that the above information is a true statement, and further state that I am not under any type of disciplinary action, including license revocation or suspension, by any State or political division thereof, or by the United States government, for any illegal or improper activity, civil or criminal, involving asbestos-containing material.

_____   _____
Applicant's Signature   Date

THIS APPLICATION MUST BE NOTARIZED
(On Page 4)

Figure 6.24 *Cont'd*

Acknowledged

County _____

State _____

Signed or affirmed before me this _____ day of _____, _____ .

_____
                                    Notary Public

My Commission Expires_____

OAC 380:50-5-8. Licensing of Asbestos Abatement Supervisors: Licensing requirements for Asbestos Abatement Supervisors are as follows: (1) Applications shall be submitted on forms prescribed by the Commissioner. (2) The license fee shall be twenty-five dollars ($25.00) per year. (3) The license shall be issued for a period not to exceed one year and shall expire concurrently with the asbestos training and subsequent refresher training. There will be no grace period wherein a Supervisor will be allowed to work with an expired license. (4) Asbestos Abatement Supervisors shall have successfully completed and shall provide documentation for: (A) an Asbestos Abatement Supervisors's course and all subsequent Supervisor refresher training that fully meets the requirements of OAC 380:50-6-3 and 380:50-6-8. (B) a two day or equivalent course in confined space entry following the NIOSH curriculum in confined space entry. (C) the NIOSH 582 course in Analysis of Airborne Asbestos Dust, or equivalent, or a minimum of a two day course in air monitoring techniques. (D) Current cardiopulmonary resuscitation (CPR) training, which may be provided by the National Heart Association, The American Red Cross, or other approved training provider. (E) Current first aid training, which may be provided by The American Red Cross, or other approved training provider. (F) Six months experience as an Asbestos Abatement Worker on job sites that have been inspected by DOL, including a minimum of six different abatement projects or containments, or one year experience as an Asbestos Abatement Worker and six months as an Asbestos Abatement Supervisor on projects that have not been inspected by DOL. (5) Licenses shall be issued in the name of the individual applicant and shall be valid only when working for a licensed Contractor. (6) License cards shall be available at the job site for inspection by the Department of Labor.

OAC 380:50-6-3. Initial training for Asbestos Contractors and Supervisors: Training requirements are as follows: (1) Training shall be supplied by an EPA or DOL accredited educational institution, labor union, or government agency, or from a private vocational education provider licensed by the state where it operates (pursuant to 70 O.S. § 21-103 within the state of Oklahoma and accredited by EPA or an EPA approved governmental agency. (2) Training providers must have EPA approval or approval by a state that has an approved Model Accreditation Plan (MAP) that is as stringent or exceeds the minimum requirements of the EPA MAP. (3) Asbestos Abatement Supervisor training courses shall be no less than five days in length and shall include: lectures, demonstrations, at least 14 hours of hands-on training, individual respirator fit testing, course review and an examination. (4) Asbestos Abatement Supervisor training courses shall consist of the training required in OAC 380:50-6-3 and 380:50-6-8.

Figure 6.24 *Cont'd*

Insulation-Containing Asbestos

7

What is vermiculite insulation? Vermiculite is a naturally occurring mineral that has the unusual property of expanding into wormlike accordion-shaped pieces when heated. The expanded vermiculite is a lightweight, fire-resistant, absorbent, and odorless material. These properties allow vermiculite to be used to make numerous products, including attic insulation.

▶ IDENTIFYING VERMICULITE INSULATION

Vermiculite can be purchased in various forms for various uses. Sizes of vermiculite products range from very fine particles to large (coarse) pieces nearly an inch long. Vermiculite attic insulation is a pebble-like, pour-in product and is usually light brown or gold in color.

Prior to its close in 1990, much of the world's supply of vermiculite came from a mine near Libby, Montana. This mine had a natural deposit of asbestos, which resulted in the vermiculite being contaminated with asbestos. Attic insulation produced using vermiculite ore, particularly ore that originated from the Libby mine, may contain asbestos fibers. Today, vermiculite is mined at three U.S. facilities and in other countries that have low levels of contamination in the finished material.

Health Problems

Asbestos can cause health problems when inhaled into the lungs. If products containing asbestos are disturbed, thin, lightweight asbestos fibers are released into the air. Persons breathing the air may breathe in asbestos fibers. Continued exposure increases the amount of fibers that remain in the lungs. Fibers embedded in lung tissue over time may result in lung diseases such as asbestosis, lung cancer, or mesothelioma. Smoking increases the risk of a person developing an illness from asbestos exposure.

Vermiculite Insulation Procedures

What should you do if you find vermiculite attic insulation? Leave it alone. You should not disturb the insulation. Any disturbance has the potential to release asbestos fibers into the air. Limit the number of trips you make into attics that contain the insulation. The Environmental Protection Agency (EPA) and Agency for Toxic Substances and Disease Registry (ATSDR) strongly recommend that

- Vermiculite insulation should be left undisturbed in attics. Due to the uncertainties with existing testing techniques, it is best to assume that the material may contain asbestos.
- People who have vermiculite insulation in their attics should not store boxes or other items in their attics if retrieving the material will disturb the insulation.
- Children should not be allowed to play in an attic with open areas of vermiculite insulation.
- Any remodeling that may disturb vermiculite should not be done before professionals who are trained and certified to handle asbestos safely remove the material.

Homeowners should make every effort to stay on the floored part of attics to avoid disturbing the insulation. If you must perform activities that may disturb the attic insulation such as moving boxes (or other materials), do so as gently as possible to minimize the disturbance. Should you disturb vermiculite insulation, leave the area immediately, unless you are properly trained and prepared to deal with asbestos exposure.

It is possible that vermiculite attic insulation can sift through cracks in a ceiling, around light fixtures, or around ceiling fans. You can prevent this problem by sealing the cracks and holes that insulation could pass through. Common dust masks are not effective against asbestos fibers. For information on the requirements for wearing a respirator mask, visit the Occupational Safety and Health Administration (OSHA) website at *http:// osha.gov/SLTC/respiratory protection/index.html*.

Testing

Not all vermiculite insulation poses a health risk due to asbestos. However, you cannot tell by looking at the insulation if it is a risk. Therefore, all precautions should be taken to avoid contact with the dust and airborne particles associated with vermiculite insulation. Essentially, you should not enter an open area containing loose insulation without asbestos removal training and protective equipment.

Collect random samples of the insulation for testing. Follow the guidelines set forth by the testing facility that you will be working with. Instructions provided by the laboratory should be followed carefully.

When you conduct your collection operation, it is important to protect all areas of a structure from exposure to the potential asbestos risk. The logical location of the insulation will be an attic. Before you open the attic access, it is wise to construct an approved barrier to prevent disturbance of the insulation from infiltrating other portions of a structure. Take your samples with as little disturbance as possible.

When the test results are available and a decision is made to remove the insulation, you will want to follow the rules and regulations set forth by OSHA and the EPA (see Chapters 8 and 9, respectively). Also check your local regulations for any amendments or differences in required procedures.

OSHA Regulations

8

The Occupational Safety and Health Administration (OSHA) has established strict exposure limits and requirements for exposure assessment, medical surveillance, recordkeeping, competent persons, regulated areas, and hazard communication. What is work classification? The OSHA standard establishes a classification system for asbestos construction work that spells out mandatory, simple, technological work practices that employers must follow to reduce worker exposures.

▶ CONSTRUCTION WORK CLASSES

Under this system, the following four classes—Class I through Class IV—of construction work are matched with increasingly stringent control requirements.

Class I Asbestos Work

Class I work is potentially the most hazardous class of asbestos jobs. This work involves the removal of asbestos-containing thermal system insulation and sprayed-on or troweled-on surfacing materials. Employers must presume that thermal system insulation and surfacing material found in pre-1981 construction is asbestos-containing material (ACM). That presumption, however, is rebuttable.

If you believe that the surfacing material or thermal system insulation is not ACM, the OSHA standard specifies the means that you must use to rebut that presumption. Thermal system insulation includes ACM applied to pipes, boilers, tanks, ducts, or other structural components to prevent heat loss or gain. Surfacing materials include decorative plaster on ceilings and walls; acoustical materials on decking, walls, and ceilings; and fireproofing on structural members.

Class II Asbestos Work

Class II work includes the removal of other types of ACM that are not thermal system insulation such as resilient flooring and roofing materials. Examples of Class II work include removal of asbestos-containing floor or ceiling tiles, siding, roofing, or transite panels.

Class III Asbestos Work

Class III work includes repair and maintenance operations where ACM or presumed ACM (PACM) are disturbed.

Class IV Asbestos Work

Class IV work includes custodial activities where employees clean up asbestos-containing waste and debris produced by construction, maintenance, or repair activities. This work involves cleaning dust-contaminated surfaces, vacuuming contaminated carpets, mopping floors, and cleaning up ACM or PACM from thermal system insulation or surfacing material.

▶ WHAT IS THE PERMISSIBLE EXPOSURE LIMIT FOR ASBESTOS?

Employers must ensure that no employee is exposed to an airborne concentration of asbestos that is in excess of 0.1 f/cc as an 8-hour time-weighted average (TWA). In addition, employees must not be exposed to an airborne concentration of asbestos in excess of 1 f/cc as averaged over a sampling period of 30 minutes.

Employers must assess all asbestos operations for the potential to generate airborne fibers, and use exposure monitoring data to assess employee exposures. They must also designate a *competent person* to help ensure the safety and health of their workers.

On all construction sites with asbestos operations, employers must designate a *competent person*—one who will be able to identify asbestos hazards in the workplace and has the authority to correct them. This person must be qualified and authorized to ensure worker safety and health as required by *Subpart C, General Safety and Health Provisions for Construction* (29 CFR Part 1926.20). Under these requirements for safety and health prevention programs, the competent person must frequently inspect job sites, materials, and equipment.

The *competent person* must attend a comprehensive training course for contractors and supervisors certified by the U.S. Environmental Protection Agency (EPA) or a state-approved training provider, or complete a course that is equivalent in length and content. For Classes III and IV asbestos work, training must include a course equivalent in length, stringency, and content to the 16-hour *Operations and Maintenance* course developed by the EPA for maintenance and custodial workers.

▶ WHAT IS AN INITIAL EXPOSURE ASSESSMENT?

To determine expected exposures, a *competent person* must perform an initial exposure assessment to assess exposures immediately before or as the operation begins. This person must perform the assessment in time to comply with all standard requirements triggered by exposure data or the lack of a negative exposure assessment and provide the necessary information to ensure all control systems are appropriate and work properly. A negative exposure assessment demonstrates that employee exposure during an operation is consistently below the permissible exposure limit (PEL).

The initial exposure assessment must be based on the following criteria:

- Results of employee exposure monitoring, unless a negative exposure assessment has been made.
- Observations, information, or calculations indicating employee exposure to asbestos, including any previous monitoring.
- For Class I asbestos work, until employers document that employees will not be exposed in excess of the 8-hour TWA PEL and short-term exposure limit (STEL), employers must assume that employee exposures are above those limits.

▶ WHAT IS A NEGATIVE EXPOSURE ASSESSMENT?

For any specific asbestos job that trained employees perform, employers may show that exposures will be below the PELs (i.e., negative exposure assessment) through the following:

- Objective data demonstrating that ACM, or activities involving it, cannot release airborne fibers in excess of the 8-hour TWA PEL or STEL.
- Exposure data obtained within the past 12 months from prior monitoring of work operations closely resembling the employer's current work operations (the work operations that were previously monitored must have been conducted by employees whose training and experience were no more extensive than that of current employees, and the data must show a high degree of certainty that employee exposures will not exceed the 8-hour TWA PEL or STEL under current conditions).

- Current initial exposure monitoring that used breathing zone air samples representing the 8-hour TWA and 30-minute short-term exposures for each employee in those operations most likely to result in exposures over the 8-hour TWA PEL for the entire asbestos job.

▶ ARE EMPLOYERS REQUIRED TO PERFORM EXPOSURE MONITORING?

Employers must determine employee exposure measurements from breathing zone air samples representing the 8-hour TWA and 30-minute short-term exposures for each employee. Employers must take one or more samples employers representing full-shift exposure to determine the 8-hour TWA exposure in each work area. To determine short-term employee exposures, employers must take one or more samples representing 30-minute exposures for the operations most likely to expose employees above the excursion limit in each work area.

Employers must also allow affected employees and their designated representatives to observe any employee exposure monitoring. When observation requires entry into a regulated area, employers you must provide and require the use of protective clothing and equipment.

When Must Employers Conduct Periodic Monitoring?

For Classes I and II jobs, employers must conduct monitoring daily that is representative of each employee working in a regulated area, unless they have produced a negative exposure assessment for the entire operation and nothing has changed. When all employees use supplied-air respirators operated in positive-pressure mode, however, employers may discontinue daily monitoring. When employees perform Class I work using control methods not recommended in the standard, employers must continue daily monitoring even when employees use supplied-air respirators.

For operations other than Classes I and II, employers must monitor all work where exposures can possibly exceed the PEL often enough to validate the exposure prediction. If periodic monitoring shows that certain employee exposures are below the 8-hour TWA PEL and the STEL, employers may discontinue monitoring these employees' exposures.

▶ IS ADDITIONAL MONITORING EVER NEEDED?

Changes in processes, control equipment, personnel, or work practices that could result in new or additional exposures above the 8-hour TWA PEL or STEL require additional monitoring regardless of a previous negative exposure assessment for a specific job.

Are Employers Required to Establish Medical Surveillance Programs?

Whether to establish medical surveillance programs for employees depends on the circumstances. Employers must provide a medical surveillance program for all employees who perform specific duties.

Engaging in Class I, II, or III work or being exposed at or above the PEL or STEL for a combined total of 30 or more days per year is one factor. Wearing negative-pressure respirators is another factor. In addition, a licensed physician must perform or supervise all medical exams and procedures that employers provide at no cost to employees and at a reasonable time. Employers must make medical exams and consultations available to employees as follows:

- Prior to employee assignment to an area where negative-pressure respirators are worn.
- Within 10 working days after the thirtieth day of combined engagement in Classes I, II, and III work and exposure at or above a PEL, and at least annually thereafter.
- When an examining physician suggests them more frequently.

If an employee was examined within the past 12 months and that exam meets the criteria of the standard, however, another medical exam is not required.

Medical Exams

When medical exams are required, they must meet specified criteria. The exams must include the following:

- Medical and work histories
- Completion of a standardized questionnaire with the initial exam and an abbreviated standardized questionnaire with annual exams
- Physical exam focusing on the pulmonary and gastrointestinal systems
- Any other exams or tests deemed necessary by the examining physician

Employers are required to provide the examining physician with the following:

- Copy of OSHA's asbestos standard and its appendices D, E, and I
- Description of the affected employee's duties relating to exposure
- Employee's representative exposure level or anticipated exposure level
- Description of any personal protective equipment and respiratory equipment used
- Information from any previous medical exams not otherwise available

It is the employer's responsibility to obtain the physician's written opinion containing results of the medical exam as well as the following information:

- Any medical conditions of the employee that increase health risks from asbestos exposure
- Any recommended limitations on the employee or protective equipment used
- A statement that the employee has been informed of the results of the medical exam and any medical conditions resulting from asbestos exposure
- A statement that the employee has been informed of the increased risk of lung cancer from the combined effect of smoking and asbestos exposure

A physician's written opinion must not reveal specific findings or diagnoses unrelated to occupational exposure to asbestos. Employers must provide a copy of the physician's written opinion to the employee involved within 30 days after receipt.

▶ EMPLOYEE RECORDS

Employers must maintain employee records concerning objective data, exposure monitoring, and medical surveillance. If using *objective data* to demonstrate that products made from or containing asbestos cannot release fibers in concentrations at or above the PEL or STEL, employers must keep an accurate record for as long as it is relied on and include the following information:

- Exempt products
- Objective data source

- Testing protocol, test results, and analysis of the material for release of asbestos
- Exempt operation and support data descriptions
- Relevant data for operations, materials, processes, or employee exposures

Employers are required to keep records of all employee *exposure monitoring* for at least 30 years, including the following information:

- Date of measurement
- Operation involving asbestos exposure that was monitored
- Methods of sampling and analysis that were used and evidence of their accuracy
- Number, duration, and results of samples taken
- Type of protective devices worn
- Name, Social Security number, and exposures of those involved

Employers are also required to make exposure records available when requested to affected employees, former employees, their designated representatives, and/or OSHA's Assistant Secretary.

In addition to retaining a copy of the information provided to the examining physician, employers must keep all *medical surveillance* records for the duration of an employee's employment plus 30 years, including the following information:

- Employee's name and Social Security number
- Employee's medical exam results, including the medical history, questionnaires, responses, test results, and physician's recommendations
- Physician's written opinions
- Employee's medical complaints related to asbestos exposure

Employers must also make employees' medical surveillance records available to them, as well as to anyone having specific written consent of an employee, and to OSHA's Assistant Secretary. Also, employers must maintain other records; for example, all employee training records for 1 year beyond the last date of employment.

If data demonstrate ACM does not contain asbestos, building owners or employers must keep associated records for as long as they rely on them. Building owners must maintain written notifications on the identification, location, and quantity of any ACM or PACM for the duration of ownership and transfer the records to successive owners.

When employers cease to do business without a successor to keep their records, employers must notify the Director of the National Institute for Occupational Safety and Health (NIOSH) at least 90 days prior to their disposal and transmit them as requested.

▶ WHAT IS A REGULATED AREA?

A regulated area is a marked-off site where employees work with asbestos, including any adjoining areas where debris and waste from asbestos work accumulate or where airborne concentrations of asbestos exceed, or can possibly exceed, the PEL.

All Classes I, II, and III asbestos work, or any other operations where airborne asbestos exceeds the PEL, must be performed within regulated areas. Only persons permitted by an employer and required by work duties to be present in regulated areas may enter a regulated area. The designated *competent person* supervises all asbestos work performed in this area.

Employers must mark off the regulated area in a manner that minimizes the number of persons within the area and protects persons outside the area from exposure to airborne asbestos. You may use critical barriers (i.e., plastic sheeting placed over all openings to the work area to prevent airborne asbestos from migrating to an adjacent area) or negative-pressure enclosures to mark off a regulated area.

Posted warning signs demarcating the area must be easily readable and understandable. The signs must bear the following information:

DANGER ASBESTOS
CANCER AND LUNG DISEASE HAZARD
AUTHORIZED PERSONNEL ONLY
RESPIRATORY AND PROTECTIVE CLOTHING
ARE REQUIRED IN THIS AREA

Employers must supply a respirator to all persons entering regulated areas. Employees must not eat, drink, smoke, chew (tobacco or gum), or apply cosmetics in regulated areas. An employer performing work in a regulated area must inform other employers onsite of the following:

- Nature of the work
- Regulated area requirements
- Measures taken to protect onsite employees

The contractor creating or controlling the source of asbestos contamination must abate the hazards. All employers with employees working near regulated areas must daily assess the enclosure's integrity or the effectiveness of control methods to prevent airborne asbestos from migrating.

General contractors on a construction project must oversee *all* asbestos work, even though they may not be the designated *competent person*. As supervisor of the entire project, the general contractor determines whether asbestos contractors comply with the standard and ensures that they correct any problems.

▶ COMMUNICATING ASBESTOS HAZARDS PRESENT AT WORK SITES

The communication of asbestos hazards is vital to prevent further overexposure. Most asbestos-related construction involves previously installed building materials. Building and facility owners often are the only or best source of information concerning these materials.

Building and facility owners, as well as employers of workers who may be exposed to asbestos hazards, have specific duties under the standard.

Before work begins, building and facility owners must identify all thermal system insulation at the work site, sprayed-on or troweled-on surfacing materials in buildings, and resilient flooring material installed before 1981. They also must notify the following persons of the presence, location, and quantity of ACM or PACM:

- Prospective employers applying or bidding for work in or adjacent to areas containing asbestos
- Building owners' employees who work in or adjacent to areas containing asbestos
- Other employers on multi-employer work sites with employees working in or adjacent to areas containing asbestos
- All tenants who will occupy the areas containing ACM

Employers discovering ACM on a work site must notify the building/facility owner and other employers onsite within 24 hours regarding its presence, location, and quantity. They also must inform owners and employees working in nearby areas of the precautions taken to confine airborne asbestos. Within 10 days of project completion, they must inform building and facility owners and other employers onsite of the current locations and quantities of remaining ACM and any final monitoring results.

At any time, employers or building and facility owners may demonstrate that a PACM does not contain asbestos by inspecting the material in accordance with the requirements of the *Asbestos Hazard Emergency Response Act* (AHERA) (40 *CFR* Part 763, Subpart E) or by performing tests of bulk samples collected in the manner described in 40 *CFR* Part 763.86.

Employers do not have to inform employees of asbestos-free building materials present; however, they must retain the information, data, and analysis supporting the determination.

OSHA-Required Warning Notices

OSHA requires the posting of warning signs (i.e., notices) when ACM or PACM risks exist. For example, a warning sign is required at the entrance to mechanical rooms or areas with ACM or PACM. The building and facility owner must post signs identifying the material present, its specific location, and appropriate work practices that ensure it is not disturbed.

Also, employers must post warning signs in regulated areas to inform employees of the dangers and necessary protective steps to take before entering.

Are Employers Required to Provide Asbestos Warning Labels?

Employers must attach warning labels to all products and containers of asbestos, including waste containers, and all installed asbestos products, when possible. Labels must be printed in large, bold letters on a contrasting background and used in accordance with OSHA's *Hazard Communication Standard* (29 *CFR* Part1910.1200).

All labels must contain a warning statement against breathing asbestos fibers and contain the following legend:

DANGER
CONTAINS ASBESTOS FIBERS
AVOID CREATING DUST
CANCER AND LUNG DISEASE HAZARD

Labels are not required if asbestos is present in concentrations less than 1 percent by weight. They also are not required if bonding agents, coatings, or binders have altered asbestos fibers, prohibiting the release of airborne asbestos over the PEL or STEL during reasonable use, handling, storage, disposal, processing, or transportation.

When building owners or employers identify previously installed asbestos or PACM, employers must attach or post clearly noticeable and readable labels or signs to inform employees which materials contain asbestos.

▶ EMPLOYEE TRAINING REGARDING ASBESTOS EXPOSURE

Employers must provide a free training program for all employees who are likely to be exposed in excess of a PEL and for all employees performing Classes I through IV asbestos operations. Employees must be trained prior to or at initial assignment and at least annually thereafter. Training courses must be easily understandable and include the following information:

- Ways to recognize asbestos
- Adverse health effects of asbestos exposure
- Relationship between smoking and asbestos in causing lung cancer
- Operations that could result in asbestos exposure and the importance of protective controls to minimize exposure
- Purpose, proper use, fitting instruction, and limitations of respirators
- Appropriate work practices for performing asbestos jobs
- Medical surveillance program requirements
- Contents of the standard
- Names, addresses, and phone numbers of public health organizations that provide information and materials or conduct smoking cessation programs
- Sign and label requirements and the meaning of their legends
- Written materials relating to employee training and self-help smoking cessation programs at no cost to employees

In addition, the following additional training requirements apply depending on the work class involved:

- For Classes I and II operations that require the use of critical barriers (or equivalent isolation methods) and/or negative-pressure enclosures, training must be equivalent in curriculum, method, and length to the EPA Model Accreditation Plan (MAP) asbestos abatement worker training.

- For employees performing Class II operations involving one generic category of building materials containing asbestos (e.g., roofing, flooring, or siding materials or transite panels), training may be covered in an 8-hour course that includes hands-on experience.

- For Class III operations, training must be equivalent in curriculum and method to the 16-hour *Operations and Maintenance* course developed by the EPA for maintenance and custodial workers whose work disturbs ACM. The course must include hands-on training on proper respirator use and work practices.

- For Class IV operations, training must be equivalent in curriculum and method to EPA awareness training. Training must focus on the locations of ACM or PACM and the ways to recognize damage and deterioration and avoid exposure.

The course must be at least 2 hours in length. Employers must provide OSHA's Assistant Secretary and the Director of NIOSH all information and training materials as requested.

▶ METHODS OF COMPLIANCE

Which methods must employers use to control asbestos exposure levels? For all covered work, employers must use the following control methods to comply with the PEL and STEL:

- Local exhaust ventilation equipped with HEPA-filter dust collection systems. A high-efficiency particulate air (HEPA) filter is capable of trapping and retaining at least 99.97 percent of all mono-dispersed particles of 0.3 micrometers in diameter.

- Enclosure or isolation of processes producing asbestos dust.

- Ventilation of the regulated area to move contaminated air away from the employees' breathing zone and toward a filtration or collection device equipped with a HEPA filter.

- Feasible engineering and work practice controls to reduce exposure to the lowest possible levels, supplemented by respirators to reach the PEL or STEL or lower.

Employers must use the following engineering controls and work practices for all operations regardless of exposure levels:

- Vacuum cleaners equipped with HEPA filters to collect all asbestos-containing or presumed asbestos-containing debris and dust.
- Wet methods or wetting agents to control employee exposures except when infeasible (e.g., due to the creation of electrical hazards, equipment malfunction, and slipping hazards).
- Prompt cleanup and disposal in leak-tight containers of asbestos-contaminated wastes and debris.

The following work practices and engineering controls are *prohibited* for all asbestos-related work or work that disturbs asbestos or PACM regardless of measured exposure levels or the results of initial exposure assessments:

- High-speed abrasive disc saws not equipped with a point-of-cut ventilator or enclosure with HEPA-filtered exhaust air.
- Compressed air to remove asbestos or ACM unless the compressed air is used with an enclosed ventilation system.
- Dry sweeping, shoveling, or other dry cleanup of dust and debris.
- Employee rotation to reduce exposure.

In addition, OSHA's asbestos standard has specific requirements for each class of asbestos work in construction.

Class I Work Compliance Requirements

A designated *competent person* must supervise all Class I work, including installing and operating the control system. The *competent person* must perform an onsite inspection at least once during each work shift and upon employee request.

Employers must place critical barriers over all openings to regulated areas or use another barrier or isolation method to prevent airborne asbestos from migrating for the following jobs:

- All Class I jobs removing more than 25 linear or 10 square feet of thermal system insulation or surfacing material.
- All other Class I jobs without a negative exposure assessment.
- All jobs where employees are working in areas adjacent to a Class I regulated area.

If using other barriers or isolation methods instead of critical barriers, employers must perform perimeter area surveillance during each work shift. No asbestos dust should be visible. Perimeter monitoring must show that clearance levels are met (as contained in 40 *CFR* Part 763, Subpart E of the *EPA Asbestos in Schools* rule) or that perimeter area levels are no greater than background levels.

Employers must ensure the following for all Class I jobs:

- Isolating heating, ventilating, and air-conditioning (HVAC) systems in regulated areas by sealing with a double layer of 6 mil plastic or the equivalent.
- Placing impermeable drop cloths on surfaces beneath all removal activity.
- Covering and securing all objects within the regulated area with impermeable drop cloths or plastic sheeting.
- Ventilating the regulated area to move the contaminated air away from the employee breathing zone and toward a HEPA filtration or collection device for jobs without a negative exposure assessment or where exposure monitoring shows the PEL is exceeded.

In addition, employees performing Class I work must use one or more of the following control methods:

- Negative-pressure enclosure systems when the configuration of the work area does not make it infeasible to erect the enclosure.
- Glove bag systems to remove ACM or PACM from piping.
- Negative-pressure glove bag systems to remove asbestos or PACM from piping.
- Negative-pressure glove box systems to remove asbestos or ACM from pipe runs.
- Water spray process systems to remove asbestos or PACM from cold-line piping if employees carrying out the process have completed a 40-hour training course on its use in addition to training required for all employees performing Class I work.
- Small walk-in enclosure that accommodates no more than two people (mini-enclosure) if the disturbance or removal can be completely contained by the enclosure.

For the specifications, limitations, and recommended work practices of these required control methods, refer to *Occupational*

Exposure to Asbestos, 29 *CFR* Part 1926.1101. Employers may use different or modified engineering and work practice controls if they adhere to the following provisions:

- The control method encloses, contains, or isolates the process or source of airborne asbestos dust, or captures and redirects the dust before it enters into the employees' breathing zone.
- A certified industrial hygienist or licensed professional engineer qualified as a project designer evaluates the work area, the projected work practices, and the engineering controls, and certifies, in writing, that based on evaluations and data, the planned control method adequately reduces direct and indirect employee exposure to or below the PEL under worst-case conditions.

The planned control method also must prevent asbestos contamination outside the regulated area, as measured by sampling meeting the requirements of the *EPA Asbestos in Schools* rule or perimeter monitoring. The employer must send a copy of the evaluation and certification to the OSHA National Office (Office of Technical Support, Room N3653, 200 Constitution Avenue, N.W., Washington, DC 20210) before using alternative methods to remove more than 25 linear or 10 square feet of thermal system insulation or surfacing material.

Class II Work Compliance Requirements

In addition to all indoor Class II jobs without a negative exposure assessment, employers must use critical barriers over all openings to the regulated area or another barrier or isolation method to prevent airborne asbestos from migrating when changing conditions indicate exposure above the PEL or when ACM is not removed substantially intact.

If using other barriers or isolation methods instead of critical barriers, employers must perform perimeter area monitoring to verify that the barrier works properly. In addition, impermeable drop cloths must cover all surfaces beneath removal activities.

All Class II asbestos work can use the same work practices and requirements as Class I asbestos jobs. Alternatively, Class II work can be performed using work practices set out in the standard for specific jobs.

For removing vinyl and asphalt flooring materials containing asbestos or installed in buildings constructed before 1981 and not verified as asbestos-free, employers must ensure that workers observe the following:

- Do not sand flooring or its backing.
- Do not rip up resilient sheeting.
- Do not dry sweep.
- Perform mechanical chipping only in a negative-pressure enclosure.
- Use vacuums equipped with HEPA filters to clean floors.
- Remove resilient sheeting by cutting with wetting of the snip point and wetting during delamination.
- Use wet methods to scrape residual adhesives and/or backing.
- Remove tiles intact, unless impossible (you may omit wetting when tiles are heated and removed intact).
- Assume resilient flooring material—including associated mastic and backing—is asbestos-containing unless an industrial hygienist determines that it is asbestos-free.

To remove asbestos-containing roofing materials, employers must ensure that workers do the following:

- Remove them intact if feasible.
- Use wet methods when intact removal is infeasible.
- Mist cutting machines continuously during use, unless the *competent person* determines misting to be unsafe.

When removing built-up roofs using a power roof cutter, employers must ensure that workers observe the following procedures:

- Use power cutters equipped with HEPA dust collectors or perform HEPA vacuuming along the cut line for roofs that have asbestos-containing roofing felts and an aggregate surface.
- Use power cutters equipped with HEPA dust collectors or perform HEPA vacuuming along the cut line, or gently sweep along the cut line and then carefully and completely wipe up the still-wet dust and debris that was acquired for roofs that have asbestos-containing roofing felts and a smooth surface.
- Do not drop or throw to the ground ACM that has been removed from a roof.
- Carry or pass the ACM to the ground by hand, or lower the material to the ground via covered, dust-tight chute, crane, or hoist.

- Lower both intact ACM and nonintact ACM to the ground as soon as it is practicable, but no later than the end of the work shift.
- Keep material wet if it is not intact or place it in impermeable waste bags or wrap it in plastic sheeting while it remains on the roof.
- Lower to the ground, as soon as possible or by the end of the work shift, any unwrapped or unbagged roofing material using a covered, dust-tight chute, crane, or hoist.
- Place unwrapped materials in closed containers to prevent scattering dust after the materials reach the ground.
- Isolate roof-level heating and ventilation air intake sources or shut down the ventilation system.

When removing cement-like asbestos-containing siding or shingles, or asbestos-containing transite panels on building exteriors other than roofs, employers must ensure that employees adhere to the following:

- Do not cut, abrade, or break siding, shingles, or transite panels unless methods less likely to result in asbestos fiber release cannot be used.
- Spray each panel or shingle with amended water before removing (amended water is water to which a surfactant [wetting agent] has been added to increase the ability of the liquid to penetrate ACM).
- Lower immediately to the ground any unwrapped or unbagged panels or shingles using a covered dust-tight chute, crane, or hoist, or place them in an impervious waste bag or wrap them in plastic sheeting and lower them to the ground no later than the end of the work shift.
- Cut nails with flat, sharp instruments.

When removing asbestos-containing gaskets, employers must ensure that employees do the following:

- Remove gaskets within glove bags if they are visibly deteriorated and unlikely to be removed intact.
- Wet the gaskets thoroughly with amended water prior to removing.
- Place the wet gaskets in a disposal container immediately.
- Keep the residue wet if removed by scraping.

For removal of any other Class II ACM, employers must ensure that employees observe the following:

- Do not cut, abrade, or break the material unless infeasible
- Wet the material thoroughly with amended water before and during removal.
- Remove the material intact, if possible.
- Bag or wrap removed ACM immediately or keep it wet until transferred to a closed receptacle no later than the end of the work shift.

Employers are allowed to use different or modified engineering and work practice controls under the following conditions:

- If they can demonstrate that employee exposure will not exceed the PEL under any anticipated circumstances.
- If a competent person evaluates the work area, the projected work practices, and the engineering controls, and certifies, in writing, that these different or modified controls will reduce all employee exposure to or below the PELs under all expected conditions of use and that they meet the requirements of the standard. This evaluation must include, and be based on, data representing employee exposure during use of the controls under conditions closely resembling those of the current job. Also, the employees participating in the evaluation must not have better training and more experience than that of the employees who are to perform the current job.

Class III Work Compliance Requirements

Employers must use wet methods and local exhaust ventilation, to the extent feasible, during Class III work. When drilling, cutting, abrading, sanding, chipping, breaking, or sawing of asbestos-containing thermal system insulation or surfacing materials occurs, employers must use impermeable drop cloths as well as mini-enclosures, glove bag systems, or other effective isolation methods and ensure that workers wear respirators.

If the material is not thermal system insulation or surfacing material and a negative exposure assessment has not been produced or monitoring shows the PEL is exceeded, employers must contain the area with impermeable drop cloths and plastic barriers or other isolation methods and ensure that employees wear respirators. In addition, the *competent person* must inspect often enough to assess changing conditions and upon employee request.

Class IV Work Compliance Requirements

Employees conducting this asbestos work must have attended an asbestos-awareness training program. They are required to use wet methods and HEPA vacuums to promptly clean asbestos-containing or presumed asbestos-containing debris. When cleaning debris and waste in regulated areas, employees must wear respirators. In areas where thermal system insulation or surfacing material is present, workers must assume that all waste and debris contain asbestos.

▶ THE COMPETENT PERSON

For Classes II, III, and IV jobs, the *competent person* must inspect often enough to assess changing conditions and upon employee request. For Class I or II asbestos work, the *competent person* must ensure the integrity of the enclosures or other containments by onsite inspection and supervise the following activities:

- Setup of regulated areas, enclosures, or other containments
- Setup procedures to control entry to and exit from the enclosure or area
- Employee exposure monitoring by ensuring it is properly conducted
- Use of required protective clothing and equipment by employees working within the enclosure or using glove bags (a plastic bag–like enclosure affixed around ACM, with glovelike appendages through which materials and tools may be handled)
- Setup, removal, and performance of engineering controls, work practices, and personal protective equipment through onsite inspection
- Use of hygiene facilities by employees
- Required decontamination procedures
- Notification requirements

► WORKING WITH RESPIRATORS

Employees must use respirators during the following activities:

- Class I asbestos jobs
- Class II work where ACM is not removed substantially intact
- Classes II and III work not using wet methods
- Classes II and III work without a negative exposure assessment
- Class III work where thermal system insulation or surfacing ACM or PACM is cut, abraded, or broken
- Class IV work within a regulated area where respirators are required
- Work where employees are exposed above the TWA or excursion limit
- Emergencies

Employers must provide respirators at no cost to workers, selecting the appropriate type from among those certified by NIOSH. Employers must provide employees performing Class I work with full-facepiece-supplied air respirators operated in pressure-demand mode and equipped with an auxiliary positive-pressure, self-contained breathing apparatus when exposure levels exceed 1 f/cc as an 8-hour TWA.

Employers must provide half-mask purifying respirators—other than disposable respirators—equipped with high-efficiency filters for Classes II and III asbestos jobs where work disturbs thermal system insulation or surfacing ACM or PACM.

If a particular job is not Class I, II, or III and exposures are above the PEL or STEL, the asbestos standard, 29 *CFR* Part 1926.1101, contains a table specifying types of respirators to use. According to 29 *CFR* Part 1910.134, employers must institute a respiratory program that includes the following:

- Procedures for selecting respirators for use in the workplace
- Fit testing procedures for tight-fitting respirators
- Procedures for proper use of respirators in routine and reasonably foreseeable emergency situations
- Procedures and schedules for cleaning, disinfecting, storing, inspecting, repairing, discarding, and maintaining respirators
- Procedures to ensure adequate air quality, quantity, and flow of breathing air for atmosphere-supplying respirators
- Training of employees in the respiratory hazards to which they are potentially exposed during routine and emergency situations

- Training of employees in the proper use and maintenance of respirators, including putting on and removing them, and any limitations on their use
- Procedures for regularly evaluating the effectiveness of the program

With regard to fit testing, employers must do the following:

- Ensure that employees are fit tested with the same make, model, style, and size of respirator that they will be using.
- Ensure that employees using a tight-fitting facepiece respirator pass an appropriate qualitative fit test (QLFT) or quantitative fit test (QNFT).
- Ensure that an employee using a tight-fitting facepiece respirator is fit tested prior to initial use of the respirator, whenever a different size, style, model, or make of respirator facepiece is used, and at least annually thereafter.
- Conduct an additional fit test whenever an employee reports (or the employer, physician, or other licensed health care professional, supervisor, or program administrator makes) visual observations of changes in an employee's physical condition that could affect respirator fit. Such conditions include, but are not limited to, facial scarring, dental changes, cosmetic surgery, or an obvious change in body weight.

Employers must not assign any employee to tasks requiring respirator use who, based on the most recent physical exam and the examining physician's recommendations, would be unable to function normally. Employers must assign such employees to other jobs or give them the opportunity to transfer to different positions in the same geographical area and with the same seniority, status, pay rate, and job benefits as they had before transferring, if such positions are available.

▶ PROTECTIVE CLOTHING

Do employers have to provide protective clothing for employees? Employers must provide and require the use of protective clothing—such as coveralls or similar whole-body clothing, head coverings, gloves, and foot coverings—for the following:

- Employees exposed to airborne asbestos exceeding the PEL or STEL
- Work without a negative exposure assessment

- Employees performing Class I work involving the removal of over 25 linear or 10 square feet of thermal system insulation or surfacing ACM or PACM

Employers must ensure that the laundering of contaminated clothing does not release airborne asbestos in excess of the PEL or STEL. Employers who give contaminated clothing to other persons for laundering must inform them of the requirement to follow procedures that do not release airborne asbestos in excess of the PEL or STEL.

Employers must transport contaminated clothing in sealed, impermeable bags or other closed impermeable containers bearing appropriate labels.

The *competent person* must examine employee worksuits at least once per work shift for rips or tears. Rips or tears found while an employee is working must be mended or the worksuit replaced immediately.

▶ CLOTHING HYGIENE-RELATED REQUIREMENTS

For this class of asbestos work, the requirements are as follows:

- Employers must create a decontamination area adjacent to and connected with the regulated area.
- Workers must enter and exit the regulated area through the decontamination area. The decontamination area must include an equipment room, shower area, and clean room in series and comply with the following:
  - Equipment room must have impermeable, labeled bags and containers to store and dispose of contaminated protective equipment.
  - Shower area must be adjacent to both the equipment and clean rooms, unless work is performed outdoors or this arrangement is not feasible (in either case, employers must ensure that employees remove asbestos contamination from their worksuits in the equipment room using a HEPA vacuum before proceeding to a shower not adjacent to the work area, or remove their contaminated worksuits in the equipment room, don clean worksuits, and proceed to a shower not adjacent to the work area).
  - Clean room must have a locker or appropriate storage container for each employee.

When it is not feasible to provide a change area adjacent to the work area, or when the work is performed outdoors, employees may clean protective clothing with a portable HEPA vacuum before leaving the regulated area. Employees then must shower and change into "street clothing" in a clean change area meeting the requirements previously described.

To enter the regulated area, employees must pass through the equipment room. But before entering the regulated area, employees must do the following:

- Enter the decontamination area through the clean room.
- Remove and deposit street clothing within a provided locker.
- Put on protective clothing and respiratory protection before leaving the clean area.

Before exiting the regulated area, employees must do at least the following:

- Remove all gross contamination and debris.
- Remove protective clothing in the equipment room (depositing the clothing in labeled, impermeable bags or containers).
- Remove respirators in the shower and then shower before entering the clean room to change into street clothing.

When workers consume food or beverages at the Class I work site, employers must provide lunch areas with airborne asbestos levels below the PEL and/or excursion limit.

▶ HYGIENE-RELATED REQUIREMENTS OF THE CLASSES

What are the hygiene-related requirements for employees performing other Class I and Classes II and III asbestos work where exposures exceed a PEL or where a negative exposure assessment has not been produced? For this class of asbestos work, the requirements are as follows:

- Employers must establish an equipment room or area adjacent to the regulated area for the decontamination of employees and their equipment.
- Workers must cover the area with an impermeable drop cloth on the floor or horizontal work surface and it must be large enough to accommodate equipment cleaning and personal protective equipment removal without spreading contamination beyond the area.

- Workers must clean the area with a HEPA vacuum before removing work clothing.
- Workers must clean all equipment and surfaces of containers filled with ACM before removal.
- Employers must ensure employees enter and exit the regulated area through the equipment room or area.

What are the hygiene-related requirements for employees performing Class IV work? For this class of asbestos work, the requirements are as follows:

- Employers must ensure that workers cleaning up dust, waste, and debris while a Class I, II, or III activity is still in progress observe the hygiene practices required of the workers performing that activity.
- Workers cleaning up asbestos-containing surfacing material or thermal system insulation debris from a Class I or III activity after the activity is finished must be provided decontamination facilities required for Class I work involving less than 25 linear or 10 square feet of material, or for Class III work where exposure exceeds a PEL or no negative exposure assessment exists.

For any class of asbestos work, employers must ensure that workers do not smoke in any work area with asbestos exposure.

▶ EMPLOYERS' HOUSEKEEPING RESPONSIBILITIES

Asbestos waste, scrap, debris, bags, containers, equipment, and contaminated clothing consigned for disposal must be collected and disposed of in sealed, labeled, impermeable bags or other closed, labeled impermeable containers. When vacuuming methods are selected, employees must use and empty HEPA-filtered vacuuming equipment carefully and in a way that will minimize asbestos reentry into the workplace.

Unless the building or facility owner demonstrates that the flooring does not contain asbestos, all vinyl and asphalt flooring material must be maintained in accordance with the following conditions:

- Sanding flooring material is prohibited.
- Employees stripping finishes must use wet methods and low-abrasion pads at speeds lower than 300 revolutions per minute.

- Burnishing or dry buffing may be done only on flooring with enough finish that the pad cannot contact flooring material.
- Employees must not dust, dry sweep, or vacuum without a HEPA filter in an area containing thermal system insulation or surfacing material or visibly deteriorated ACM.
- Employees must promptly clean up the waste and debris and accompanying dust, and dispose of it in leak-tight, labeled containers.

▶ OSHA ASSISTANCE

OSHA can provide extensive help through a variety of programs, including technical assistance about effective safety and health programs, state plans, workplace consultations, voluntary protection programs, strategic partnerships, training and education, and more. An overall commitment to workplace safety and health can add value to a business, to the workplace, and to people's lives.

What Are the Safety and Health System Management Guidelines?

Effective management of worker safety and health protection is a decisive factor in reducing the extent and severity of work-related injuries and illnesses and their related costs. In fact, an effective safety and health program forms the basis of good worker protection and can save time and money—about $4 for every dollar spent—and increase productivity and reduce worker injuries, illnesses, and related worker compensation costs.

To assist employers and employees in developing effective safety and health programs, OSHA published recommended *Safety and Health Program Management Guidelines*. These voluntary guidelines can be applied to all places of employment covered by OSHA.

The guidelines identify four general elements critical to the development of a successful safety and health management system:

- Management leadership and employee involvement
- Work site analysis
- Hazard prevention and control
- Safety and health training

The guidelines recommend specific actions, under each of these general elements, to achieve an effective safety and health program. The Federal Register notice is available online at *www. osha.gov*.

What Are State Programs?

The *Occupational Safety and Health Act of 1970* (OSH Act) encourages states to develop and operate their own job safety and health plans. OSHA approves and monitors these plans. There are currently 26 state plans. There are 23 that cover both private and public (state and local government) employment. Three states—Connecticut, New Jersey, and New York—cover the public sector only. States and territories with their own OSHA-approved occupational safety and health plans must adopt standards identical to, or at least as effective as, the federal standards.

How to Obtain Consultation Services

Consultation assistance is available on request to employers that want help in establishing and maintaining a safe and healthful workplace. Largely funded by OSHA, the service is provided at no cost to the employer. Primarily developed for smaller employers with more hazardous operations, the consultation service is delivered by state governments employing professional safety and health consultants. Comprehensive assistance includes an appraisal of all mechanical systems, work practices, and occupational safety and health hazards of the workplace, and all aspects of the employer's present job safety and health program.

In addition, the service offers assistance to employers in developing and implementing an effective safety and health program. No penalties are proposed or citations issued for hazards identified by the consultant. OSHA provides consultation assistance to the employer with the assurance that the owner's name and firm and any information about the workplace will not be routinely reported to OSHA enforcement staff.

Under the consultation program, certain exemplary employers may request participation in OSHA's Safety and Health Achievement Recognition Program (SHARP). Eligibility for participation in SHARP includes receiving a comprehensive consultation visit, demonstrating exemplary achievements in workplace safety and health by abating all identified hazards, and developing an excellent safety and health program.

Employers accepted into SHARP may receive an exemption from programmed inspections (not complaint or accident investigation inspections) for a period of 1 year.

For more information concerning consultation assistance, see the OSHA web site at *www.osha.gov*.

What Are Voluntary Protection Programs?

Voluntary Protection Programs (VPPs) and onsite consultation services, when coupled with an effective enforcement program, expand worker protection to help meet the goals of the OSH Act. The three VPPs—Star, Merit, and Demonstration—are designed to recognize outstanding achievements by companies that have successfully incorporated comprehensive safety and health programs into their total management system. The VPPs motivate others to achieve excellent safety and health results in the same outstanding way as they establish a cooperative relationship between employers, employees, and OSHA.

For additional information on VPPs and how to apply, contact the OSHA regional offices.

What Is the Strategic Partnership Program?

OSHA's Strategic Partnership Program, the newest member of OSHA's cooperative programs, helps encourage, assist, and recognize the efforts of partners to eliminate serious workplace hazards and achieve a high level of worker safety and health. Whereas OSHA's Consultation Program and VPP entail one-on-one relationships between OSHA and individual work sites, most strategic partnerships seek to have a broader impact by building cooperative relationships with groups of employers and employees. These partnerships are voluntary, cooperative relationships between OSHA, employers, employee representatives, and others (e.g., trade unions, trade and professional associations, universities, and other government agencies).

For more information on this program, contact your nearest OSHA office or visit OSHA's web site at *www.osha.gov*.

Does OSHA Offer Training and Education?

OSHA's area offices offer a variety of information services, such as compliance assistance, technical advice, publications, audiovisual aids, and speakers for special engagements. OSHA's Training Institute in Des Plaines, Illinois, provides basic and advanced courses in safety and health for federal and state compliance officers; state consultants; federal agency personnel; and private sector employers, employees, and their representatives.

The OSHA Training Institute also has established OSHA Training Institute Education Centers to address the increased demand for its courses from the private sector and from other federal agencies. These centers are nonprofit colleges, universities, and other organizations that have been selected after a competition for participation in the program.

OSHA also provides funds to nonprofit organizations, through grants, to conduct workplace training and education in subjects where OSHA believes there is a lack of workplace training. Grants are awarded annually. Grant recipients are expected to contribute 20 percent of the total grant cost.

For more information on grants, training, and education, contact the OSHA Training Institute, Office of Training and Education, 1555 Times Drive, Des Plaines, IL 60018; telephone: 847-297-4810. For further information on any OSHA program, contact your nearest OSHA area or regional office.

Does OSHA Provide Any Information Electronically?

OSHA has a variety of materials and tools available on its web site at *www.osha.gov*. These include eTools such as Expert Advisors, Electronic Compliance Assistance Tools (e-CATs), technical links, regulations, directives, publications, videos, and other information for employers and employees.

OSHA's software programs and compliance assistance tools walk you through challenging safety and health issues and common problems to find the best solutions for the workplace.

The Occupational Safety and Health Administration's online help includes standards, interpretations, directives, and more— see *www.osha.gov/dts/osta/oshasoft/asbestos/index.html*. For publications, contact the U.S. Government Printing Office, Superintendent of Documents, P.O. Box 371954, Pittsburgh, PA 15250-7954; telephone: 202-512-1800.

How Do I Learn More About Related OSHA Publications?

OSHA has an extensive publications program. For a listing of free or sales items, visit OSHA's web site at *www.osha.gov* or contact the OSHA Publications Office, U.S. Department of

Labor, 200 Constitution Avenue, N.W., N-3101, Washington, DC 20210; telephone: 202-693-1888; or fax to 202-693-2498.

How Do I Contact OSHA?

To report an emergency, file a complaint, or seek OSHA advice, assistance, or products, call 1-800-321-OSHA or contact your nearest OSHA regional or area office. The teletypewriter (TTY) number is 1-877-889-5627.

You can also file a complaint online and obtain more information on OSHA federal and state programs by visiting OSHA's web site at *www.osha.gov*. For more information on grants, training, and education, contact the OSHA Training Institute, or see Outreach on OSHA's web site.

► CONCLUSION

This chapter completes our overview of OSHA as it pertains to asbestos. OSHA and the EPA have set forth a magnitude of information and regulations for asbestos workers. With OSHA now behind us, let's move on to the EPA guidelines.

EPA Requirements and Recommendations

9

The U.S. Environmental Protection Agency (EPA) requirements and recommendations for working with asbestos are vast. They make the Occupational Safety and Health Administration (OSHA) regulations look short in comparison. Both the OSHA and the EPA rulings must be followed when working with asbestos. I found a revised set of EPA regulations to pull from for this chapter. The regulations were revised in April 2011. These particular rules apply to the state of Maine, where I live and work. I chose this material due to its recent nature. If you are ready for a long read, let's get started.

▶ DEFINITIONS

The Environmental Protection Agency sets forth its own definitions. To make sense of the rules, you have to know the definitions, so we will start with them.

Abrade: To wear away or rub off by friction asbestos-containing materials such that the asbestos-containing material is friable.

ACM: Asbestos-containing material.

Adequately wet: To sufficiently mix or penetrate with liquid to prevent visible emissions and the release of any particulates during handling. Asbestos-containing material that is adequately wet will have no visible emissions when handled and will feel moist to the touch. The absence of visible emissions is not sufficient evidence of being adequately wet.

Aggressive method: Removal or disturbance by sanding, cutting, grinding, or abrading.

Aggressive sampling: Sweeping the walls, ceiling, and floor of a regulated area with the exhaust of a minimum one-horsepower leaf blower immediately preceding air clearance sampling, then placing a stationary fan at a shallow angle to the floor for each 10,000 cubic feet of regulated area with the fan air directed to the ceiling, and running the fan(s) throughout the air sampling event.

AHERA: Asbestos Hazard Emergency Response Act enabling legislation, enacted on October 22, 1986, authorizing EPA to promulgate the "Asbestos-Containing Materials in Schools" rule, 40 CFR, Part 763 [as amended], with an effective date of December 14, 1987.

AIHA: American Industrial Hygiene Association.

Air clearance sampling: Air monitoring conducted by a certified Asbestos Air Monitor at the conclusion of an asbestos abatement activity.

Air monitoring: Collecting samples of air before, during, or after an asbestos abatement activity to measure the concentration of airborne fibers.

Alter: To modify, change, or remove.

Applicant: An individual, business entity, or public entity formally requesting from the Department a license or certificate to engage in an asbestos abatement activity regulated by this rule.

Area monitoring: Air monitoring, excluding personal sampling, performed inside, outside, and/or adjacent to the regulated area to determine whether elevated fiber counts are being generated during a particular abatement activity.

Asbestos: A group of naturally occurring minerals that separate into fibers of high tensile strength and are resistant to heat, wear, and chemicals, including but not limited to the asbestiform varieties of serpentine (chrysotile), riebeckite (crocidolite), cummingtonitegunerite (amosite), anthophyllite, actinolite, tremolite, and any of these minerals that have been chemically treated or altered.

Asbestos abatement activity: For purposes of this rule, this means any activity involving the removal, demolition, enclosure, repair, encapsulation, or handling of asbestos-containing materials in an amount greater than 3 square feet or 3 linear feet. It includes associated activities such as design, monitoring, analysis, and inspection of asbestos-containing materials in an amount greater than 3 square feet or 3 linear feet, and conducting training for persons seeking a state certificate or license.

Asbestos abatement contractor: A business entity licensed by the Department that engages in, or intends to engage in, asbestos abatement activities as a business service and that employs or involves one or more asbestos abatement project supervisors, asbestos abatement workers, Asbestos Abatement Design Consultants, Asbestos Air Monitors, or Asbestos Inspectors for asbestos abatement activities.

Asbestos Abatement Design Consultant: A Department-certified individual engaged in preparing and supervising the implementation of facility plans for the removal or abatement of asbestos. These activities include, but are not limited to, the performance of air quality and bulk sampling; advising building owners, contractors, and project supervisors on health impacts of asbestos abatement activities; and supervising the conduct of training courses. This category of specialists includes, but is not limited to, engineers, architects, health professionals, industrial hygienists, private consultants, or other individuals involved in asbestos risk assessment or regulatory activities.

Asbestos Abatement Project Supervisor: A Department-certified individual with responsibility for the supervision of asbestos abatement activities. These persons include, but are not limited to, project supervisors employed by contractors or by in-house Asbestos Abatement Units; and employees of governmental or public entities who coordinate or directly supervise asbestos abatement activities performed by public schools, governmental, or other public employees in a school district, governmental, or other public buildings.

Asbestos Abatement Worker: A Department-certified individual engaging in any asbestos abatement activity for any employer.

Asbestos Air Analyst: A Department-certified individual engaging in the analysis of air samples for fiber count including but not limited to asbestos.

Asbestos Air Monitor: A Department-certified individual who is responsible for conducting air and/or project monitoring prescribed by this rule and other rules and regulations to protect the public health from the hazards associated with exposure to asbestos.

Asbestos Analytical Laboratory: A business entity or public entity licensed by the Department that qualitatively or quantitatively analyzes samples of solids, liquids, or gases for fibers including asbestos fibers.

Asbestos-associated activity: "Asbestos-associated activity" means asbestos-related activity, such as inspection, design, monitoring, analysis, and the provision of training to persons seeking to become certified as asbestos professionals, conducted generally in support of asbestos abatement activity, but excluding removal, demolition, enclosure, handling, repair, or encapsulation of ACM.

Asbestos Bulk Analyst: A Department-certified individual engaging in the analysis of bulk samples for asbestos and/or other material composition for qualitative or quantitative purposes.

Asbestos Consultant: A business entity licensed by the Department that engages in, or intends to engage in, the design, inspection, or monitoring of asbestos abatement activities.

Asbestos-containing cementitious product: Any preformed material or product manufactured with asbestos mixed into cement. Commonly referred to as transite, it includes but is not limited to piping, board, and siding.

Asbestos-containing material: Any material containing asbestos in quantities greater than or equal to 1% by volume as determined by weight, visual evaluation, and/or point count analysis.

Asbestos-contaminated material: Any solid, liquid, or waste material that contains asbestos or is contaminated with asbestos.

Asbestos impact survey: Asbestos inspection.

Asbestos Inspector: A Department-certified individual whose activities include, but are not limited to, collecting bulk samples and assessing the potential for exposure associated with the presence of ACM.

Asbestos management planner: A Department-certified individual who assesses hazards associated with the presence and condition of ACM in schools, and who develops a response action plan based upon assessment.

Asbestos professional: An individual certified by the Department to engage in asbestos abatement activities, including but not limited to Asbestos Abatement Worker, Asbestos Abatement Project Supervisor, Asbestos Air Monitor, Asbestos Inspector, Asbestos Abatement Design Consultant, Asbestos Management Planner, Asbestos Bulk Analyst, and Asbestos Air Analyst.

Asbestos waste: Any asbestos-contaminated material, asbestos debris, or asbestos-containing waste generated from an asbestos abatement activity or any other source that is discarded and considered a waste material.

Asbestos waste storage facility or AWSF: A temporary storage container or building where more than one cubic yard of asbestos waste is stored. A AWSF does not include the storage of asbestos waste on the site of generation during the asbestos abatement activity and for a period not exceeding 5 days after completion of the project or areas where asbestos waste is being stored or placed for less than 1 day, such as vehicles, loading docks, and staging areas.

Board: The Maine Board of Environmental Protection.

Building: Any discrete structure suitable for housing individuals, equipment, or other items, including but not limited to the heating, ventilation, and air-conditioning systems servicing that building. The foundation layout generally comprises the dimensions of a "building." For purposes of this rule, structures connected by a portico, exterior hallway, corridor, or similar passageway are not considered to be the same "building."

Business entity: A partnership, firm, association, corporation, sole proprietorship, or any other form of business concern.

Certificate: A document issued to an individual by the Commissioner affirming that an individual has successfully completed the training and other requirements set forth in this rule to qualify as an asbestos professional.

Clean room: The section of a decontamination facility or unit where clean clothes and towels are located and that separates the shower room from the outside of a containment or regulated area. The clean room must be a minimum of 24 square feet for projects involving three or fewer certified individuals working for the contractor entering the regulated area per work shift and a minimum of 32 square feet for all other projects.

Commissioner: The Commissioner of the Department of Environmental Protection.

Competent person: For roofing, cementitious products, and demolition by large equipment projects subject to OSHA jurisdiction, the individual capable of identifying existing asbestos hazards in the work site who has the authority to take prompt corrective measures to eliminate such hazards, as defined in the OSHA Asbestos Standard for Construction, 29 CFR 1926.1101 (amended and corrected), effective date August 10, 1994. Note: A competent person must attend a comprehensive asbestos training course (i.e., the asbestos abatement supervisor training course set forth in EPA's Model Accreditation Program [40 CFR, Part 763, Appendix C to Subpart E] conducted by an accredited asbestos Training Provider, or a course that is equivalent in length and content).

Containerize: The sealing of asbestos waste in fiber-tight polyethylene sheeting; in a fiber-tight metal, plastic, or fiber drum with a locking lid; or in leak-proof containers.

Containment: A specific area designated for asbestos abatement activity in which engineering control measures have been implemented to prevent the release of asbestos fibers to the atmosphere by means of covering the walls, ceilings, and floors, and openings of such areas with polyethylene sheeting; establishing reduced atmospheric pressure within the area; and controlling personnel access to the area.

Critical barrier: A fiber-tight barrier within 15 feet of the regulated area and within the room(s) in which the project is conducted consisting of one layer of 6-mil polyethylene sheeting, that is a separate layer from containment, used to seal windows (excluding fixed windows), doors, vents, drains, wall penetrations, and any other penetrations.

Cut: To penetrate with a sharp-edged instrument, including sawing, but not including shearing, slicing, or punching.

Decontamination facility or unit: An enclosed area adjacent and connected to the containment consisting of, at a minimum, an equipment room, a shower room, and a clean room that is used for the decontamination of workers, materials, and equipment.

Demolition: The act or process of tearing down or razing a building or discrete portion thereof together with any associated handling operations, or the intentional burning of a building.

Department: The Maine Department of Environmental Protection composed of the Board and the Commissioner.

Deteriorated: The condition of a material in which the binding or matrix is losing or has lost its integrity so that the material is friable.

Emergency asbestos abatement activity: An asbestos abatement activity necessitated by a sudden, unexpected event such as nonroutine failures of equipment or by actions of fire and emergency medical personnel pursuant to duties within their official capacities.

Employee: Each person who may be permitted, required, or directed by an employer in consideration of direct or indirect gain or profit to engage in any employment.

Employer: An individual, business entity, or public entity that gives an employee, whether directly or indirectly, gain or profit in exchange for performing any task.

Encapsulant: Liquid sealant applied to ACM to reduce the tendency of ACM to release asbestos fibers.

Encapsulation: The application of a liquid sealant to ACM or the wrapping of ACM with rewettable cloth and mastic to reduce the tendency of the ACM to release asbestos fibers. Except for friable-surfacing ACM, painting or sealing intact ACM to prevent damage or to enhance appearance is not an asbestos abatement (encapsulation) activity as long as painting or sealing intact ACM will not disturb or damage the ACM.

Enclosure: The covering of ACM in, under, or behind any kind of fiber-tight barrier such as walls. Enclosing intact ACM to prevent damage or to enhance appearance is not asbestos abatement (enclosure) activity as long as enclosing the intact ACM will not disturb or damage the ACM.

Equipment room: An asbestos-contaminated room located within the decontamination facility or unit that is connected to both the containment and the shower, and supplied with impermeable bags or containers for the disposal of contaminated protective clothing, equipment, and asbestos waste. The equipment room may or may not be part of a designated waste load-out unit.

Facility: An institutional, commercial, public, industrial, or residential structure, installation, or building or set of buildings at a single geographic location, including those containing condominium units, or individual dwelling units operated as a residential cooperative, military or company housing, or ship.

Friable ACM or friable asbestos: Any ACM that, when dry, has the potential to readily release asbestos fibers when crumbled, pulverized, handled, deteriorated, or subjected to mechanical, physical, or chemical processes. It also means potentially friable ACM that has deteriorated or has been or will be processed to the extent that, when dry, it may readily release asbestos fibers. Sanding, cutting, abrading, or grinding are processes that will readily release asbestos fibers from potentially friable ACM. Activities that render potentially friable asbestos-containing materials friable include, but are not limited to, removal of asbestos flooring by an aggressive method, and subjecting asbestos-containing cementitious products and asbestos-containing latex, asphaltic, or petroleum-based materials to sanding, grinding, abrading, or cutting with a mechanical cutter or to mechanical and/or physical forces such that the material is no longer intact.

Glove bag: A polyethylene bag, minimum 6 mil in thickness or equivalent, with built-in gloves that is used to remove ACM in small quantities.

Grind: To reduce to powder or small fragments using such methods as, but not limited to, mechanical chipping, grinding, drilling, or shot or bead blasting.

Gross visible debris: Visible debris, other than dust, in the regulated area.

Handling: To lift, impact, dislodge, process, remove, store, or otherwise manipulate friable ACM.

HEPA: A high-efficiency particulate air filter capable of retaining 0.3-micrometer-diameter particles with 99.97% efficiency, including N-100, R-100, or P-100 filter cartridges.

Homogenous area: A discrete portion of surfacing material, thermal system insulation, or miscellaneous ACM that is uniform in color, texture, and composition.

In-house Asbestos Abatement Unit: An employee or group of employees of a Department-licensed business or public entity that engages in, or intends to engage in, asbestos abatement activities at projects solely within the confines of property owned or leased by the entity and that employs one or more asbestos abatement supervisors or other asbestos professionals.

Independent business relationship: A relationship between two businesses in which no financial or shareholder control is exerted by one over the other, except by contract to perform services for a specific project where the relationship between the contracting businesses is otherwise independent and the facility owner or agent is aware of the contractual relationship. Circumstances where a relationship between two businesses is not independent include but are not limited to:

- When a person and/or immediate family member or business entity has ownership shares in both businesses.
- When a person and/or immediate family member has ownership in, or serves as an officer, director, or employee of, one business, and serves as an officer, director, or employee of another business.

- When a person or business entity with ownership in, or serving as an officer, director, or employee of, one business has provided capital or other financial support to another business.

When a person or business entity hires another person or business entity to conduct visual evaluation(s) and/or air clearance sampling using a fixed-price contract that does not fully compensate for additional sampling, analytical, and labor costs necessitated by failure of the visual evaluation and/or air clearance sampling. Independent business relationship is defined for purposes of the independence requirements in Sections 2(G), 4(E)(4)(a), 5(C)(1)(c), and 8(A)(1) of this rule and the financial-interest disclosure requirements in Sections 6(D)(1) and 6(D)(2).

Inspection: The process of visually identifying the locations, collecting bulk samples, and assessing the condition of suspect ACM present in or around a facility. This process includes sampling and assessing suspect ACM and creating records documenting the same. Inspection includes conducting a comprehensive building survey by a certified Asbestos Inspector prior to renovation and/or demolition activities to determine the presence of any asbestos-containing materials that may be impacted by the renovation and/or demolition project.

Intact: The ACM has not crumbled, been pulverized, or otherwise deteriorated so that the asbestos is no longer likely to be bound with its matrix.

License: A document issued by the Commissioner to a business entity or public entity affirming that the entity has met the requirements set forth in this rule to engage in asbestos abatement activities including but not limited to Asbestos Abatement Contractor, in-house Asbestos Abatement Unit, Asbestos Consultant, Asbestos Analytical Laboratory, and Training Provider.

Maine asbestos law: Maine Revised Statutes Title 38, Sections 1271 to 1284.

Mastic: Petroleum- or asphalt-based adhesive, glue, or protective coating used to adhere or fasten a product to a surface or to protect a surface from corrosion or decay. As used in this rule, mastic does not include surfacing materials or thermal system insulation.

Mechanical chipping: Grinding by machine or tool, including but not limited to ice scraper, spud bar, pry bar, and mechanical chipping machine.

Miscellaneous ACM: All ACM that is not surfacing material or thermal system insulation.

Model Accreditation Plan (MAP): Appendix C, the Interim Final Rule-59 FR 5236-5260 (effective April 4, 1994) to Subpart E of the "Asbestos-Containing Materials in Schools" rule, 40 CFR, Part 763 (effective December 14, 1987).

Nonfriable ACM: ACM that is not friable ACM and is intact and not deteriorated. Nonfriable ACM generally contains asbestos that is mixed into another medium or matrix such as vinyl or cement, and does not emit fibers readily unless deteriorated or subjected to sanding, cutting, grinding, or abrading.

NVLAP: National Voluntary Laboratory Accreditation Program.

OSHA: The federal Occupational Safety and Health Administration.

Owner or Operator: Any person who owns, leases, operates, controls, or supervises an asbestos abatement activity within a facility or who owns, leases, operates, controls, or supervises the facility at which an asbestos abatement activity occurs, or both.

PAT: Proficiency Analytical Testing, prescribed by AIHA to determine Asbestos Analytical Laboratory or analyst proficiency.

Person: Any individual, business entity, governmental body, or other public or private entity.

Physical barrier: A partition utilized to prevent access into a regulated area by a person not associated with the project.

Potentially friable: Intact ACM that is not currently friable but will become friable if it is sanded, grinded, abraded, or cut with a mechanical cutter or subject to mechanical/physical forces such that it is no longer intact.

Project: The asbestos abatement activities outlined in a project design, including any precleaning and the hanging of polyethylene sheeting, that occur within a single building during a discrete and finite time period and with a lapse in abatement activity of no more than 10 working days.

Project completion: The end of all asbestos abatement activities in a regulated area.

Project design: An asbestos-specific plan or any related set of directions, work orders, bid documents, or specifications developed by an Asbestos Abatement Design Consultant for the removal, enclosure, encapsulation handling, or repair of more than 3 square or 3 linear feet of ACM, or for the demolition or renovation of a facility or facility component that contains or impacts more than 3 square or 3 linear feet of ACM.

Project monitoring: An asbestos-associated activity undertaken by an Asbestos Air Monitor that documents whether an asbestos abatement activity is conducted and completed according to design specifications and applicable rules and regulations, or that performs area monitoring inside, outside, and/or adjacent to the regulated area to determine whether elevated fiber counts are being generated during a particular abatement activity.

Public entity: The state, any of its political subdivisions, or any agency or instrumentality of either.

Regulated area: An area established by the owner or operator to demarcate the geographic boundaries where asbestos abatement activities take place. Whenever a work area containment is used, it demarcates the regulated area. Establishing the regulated area includes demarcation, precleaning, hanging polyethylene sheeting, and any other activities that have the potential to disturb asbestos-containing materials at the site.

Remote decontamination facility or unit: A decontamination facility or unit that is not contiguous with the regulated area, where negative air pressure is not required.

Removal: Taking out ACM or facility components that contain or are covered with ACM from a facility.

Renovation: The removal of any asbestos-containing facility component(s) and/or building materials together with any related handling operations.

Repair: An asbestos abatement activity that involves sealing, patching, and/or enclosure of damaged ACM.

Responsible person: An individual having ultimate legal control of actions of a business entity, public entity, or employer, or such individual's agent.

Sampling: The process of obtaining representative portions of materials or air suspected to contain asbestos, including but not limited to the taking of bulk samples of materials, the collection of liquids, or the collection of air for the purposes of enumerating asbestos or fiber concentrations.

Sand: To polish or rub off with sandpaper.

Shower room: A section of the decontamination facility or unit that has a shower equipped with continuous running, adjustable, hot and cold water; soap; and a mechanism of containing and collecting shower water that must be filtered with a 5-micron filter prior to discharge.

State: The State of Maine.

Static Air Clearance Sample: An air clearance sample that is collected from a regulated area that is not subjected to aggressive sampling.

Surfacing material: ACM that was sprayed on, troweled on, or otherwise applied to surfaces for acoustical, fireproofing, or other purposes.

TEM: Transmission electron microscopy.

Thermal system insulation: ACM that was applied to pipes, ducts, fittings, boilers, breaching, tanks, or other components to prevent heat loss or gain, or water condensation, or for other purposes.

Training Provider: A person licensed by the Department to provide training that is necessary to fulfill certification or licensing requirements under this rule.

U.S. EPA: U.S. Environmental Protection Agency.

Visible debris: Any particulate residue, such as dust, dirt, or other extraneous material, that may or may not contain asbestos, capable of detection by the human eye without the aid of instruments.

Visible emission: Any emission originating from ACM during abatement activities and capable of being detected by the human eye without the aid of instruments.

Visual evaluation: The process by which the regulated area is scrutinized by a certified Asbestos Air Monitor to ascertain whether visible debris is present upon completion of an asbestos abatement activity.

Waste disposal site: A land area, facility, location, or combination of them, including landfills, utilized for final disposal of asbestos waste.

Waste load-out unit: Part of a containment through which asbestos waste or asbestos-contaminated material or supplies are removed from the regulated area. Waste load-out units shall be equipped with an air-lock contiguous with the containment to allow cleaning and packaging of waste.

Work area: Any physical space within the regulated area in which an asbestos abatement activity is being performed.

Work practices: The minimum standards, procedures, or actions taken or used in carrying out any asbestos abatement activity.

Work site: The geographic location, as indicated by street address, at which an asbestos abatement activity takes place.

► GENERAL PROVISIONS

The rules and regulations that follow apply to asbestos abatement activities, including removal, encapsulation, demolition, enclosure, repair, and handling, and associated activities such as inspection, design, analysis, monitoring, and training, conducted in the State of Maine. Disposal of asbestos and asbestos-containing material is governed by Maine's *Landfill Siting, Design and Operation Rule,* 06-096 CMR 401. Transportation of asbestos or asbestos-containing waste material is governed by Maine's *Transportation of Hazardous Materials Rule,* 16-219 CMR 60 (effective July 27, 2010, as amended), as administered by the Maine Department of Public Safety.

Employees and agents of the Department may enter any property at reasonable hours and enter any building with the consent of the property owner or operator, occupant or agent, or pursuant to an administrative search warrant, to inspect the property or structure, take samples, or conduct tests as appropriate to determine compliance with any asbestos laws and regulations administered by the Department or the terms and conditions of any order, license, permit, approval, or decision of the Commissioner or the Board.

No person or owner or operator may engage in, or arrange for, any asbestos abatement activities in Maine unless the asbestos abatement activity is performed by licensed entities and certified professionals, notification is provided to the Department, and the appropriate work practice standards are followed, in accordance with this rule.

Exemptions

The disposal of asbestos waste or other asbestos abatement activity related to disposal at a site licensed to accept asbestos waste for disposal is exempt from this rule. Persons undertaking asbestos abatement activities in single-unit residential buildings are exempt from the licensing and certification requirements of these rules provided that the activities are limited to heating equipment and performed by persons licensed by the Oil and Solid Fuel

Board under Title 32, chapter 33, to install, repair, remove, or service heating equipment. These persons must comply with the notification, work practices, and other requirements of these rules.

Activities Not Subject to This Rule

The following activities are not subject to this rule. While the activities delineated next are not subject to this rule, they are subject to the following regulations:

- Federal OSHA General and Construction Standards apply to all of the removal/containerization activities listed next; removal requirements include but are not limited to work practice and engineering controls. Containerization requirements include placing asbestos waste in leak-proof containers. OSHA regulations apply whenever a business entity employs individuals for compensation (homeowners are not subject to OSHA's regulations). Please contact OSHA for information on how to comply with their standards.

- The Federal Asbestos in Schools Rule (commonly referred to as AHERA) contains specific requirements for all asbestos-related activities that take place in public schools including but not limited to training requirements for school personnel who disturb asbestos. School personnel need to check with the school's designated person before undertaking any asbestos-related activities.

- The transportation of asbestos-containing materials is governed by Maine's Non-Hazardous Waste Transporter licenses, 06-096 CMR 411. This rule requires that businesses that transport asbestos-containing material be licensed by the Department before doing so.

- The disposal of asbestos-containing materials in Maine is governed by Maine's Landfill Siting, Design, and Operation Rule, 06-096 CMR 401. Please contact the DEP Asbestos Hazard Prevention Program at 207-287-2651 to receive information on how to perform the activities listed next safely and how not to become subject to the requirements of this rule.

  (1) The removal and containerization of intact asbestos-containing latex, asphaltic, or petroleum-based materials including but not limited to mastics, glues, cements, sealants, coatings, and adhesives provided they are not sanded, grinded, abraded, or cut with a mechanical cutter.

The use of bead or shot blasting equipment constitutes grinding and is therefore a regulated activity. Scraping and scrubbing asbestos-containing asphaltic or petroleum-based materials is not considered sanding, cutting, grinding, or abrading.

Note: The work practice requirements for the removal and containerization of greater than 3 square feet of asbestos-containing latex, asphaltic, or petroleum-based materials that are sanded, grinded, or abraded are set forth in Section 7(D) of this rule. The work practice requirements for the removal and containerization of more than 105 square feet of asbestos-containing latex, asphaltic, or petroleum-based materials that are cut with mechanical roof cutters are set forth in Section 7(D); performing these activities requires licensure as an asbestos abatement contractor. The 105-square-feet threshold is based on EPA's "Interpretive Rule Governing Roof Removal Operations, 40 CFR, Part 61, Appendix A to Subpart M, dated June 17, 1994, which states that cutting $5,580\,ft^2$ of asbestos-containing roofing material creates $160\,ft^2$ of friable ACM, the point at which asbestos-containing roof projects become subject to federal NESHAP requirements; cutting $105\,ft^2$ of asbestos-containing roofing material creates $3\,ft^2$ of ACM, the regulatory threshold of an "asbestos abatement activity" per Maine asbestos laws and regulations.

(2) The removal and containerization of asbestos-containing cementitious products such as exterior siding at owner-occupied single-family residential units when performed by the homeowner.

Homeowners should contact the DEP Asbestos Hazard Prevention Program at 207-287-2651 to receive information on how to safely perform this activity. If you live in a single-family home that you own and want someone else to remove your asbestos siding for you, you must hire a licensed asbestos abatement contractor. The removal of asbestos siding from multifamily dwellings and commercial buildings must be done by a Maine-licensed asbestos abatement contractor.

(3) The removal and containerization of intact asbestos-containing cementitious piping and electrical conduits

provided they are not sanded, grinded, abraded, or cut with a mechanical cutter. Each section must be removed using best management practices such that a minimum amount of breakage occurs during the initial removal of fasteners, other attaching system or decoupling, and the product remains intact throughout the remainder of the removal and containerization process.

The work practice requirements for the removal and containerization of greater than 3 square feet of other asbestos-containing cementitious materials not listed before and for asbestos-containing piping and electrical conduits that are sanded, grinded, abraded, or cut with a mechanical cutter are set forth in Section 7(D) of this rule; performing these activities requires licensure as an asbestos abatement contractor.

(4) The removal and containerization of intact asbestos-containing caulking or glazing. Note: The work practice requirements for the removal and containerization of intact asbestos-containing caulking or glazing that are sanded, grinded, abraded, or cut with a mechanical cutter are set forth in Section 7(A) of this rule; performing these activities requires licensure as an asbestos abatement contractor.

(5) The removal and containerization of asbestos-containing joint compound used to fill nail holes and tape seams in building component systems including but not limited to walls and ceilings. Note: Joint compound used as a layered system on walls, ceilings, etc. is not exempt.

(6) The removal and containerization of asbestos-containing gaskets when the gasket itself is encased by a facility component and will not be disturbed or impacted by the operation. Note: An example of this type of operation would be removal of a flange gasket by cutting or removing the piping on either side of the flange.

(7) The removal and containerization of intact asbestos-containing flooring except for felt-backed sheet flooring products, in a nonaggressive method, using infrared tile lift machine or heat guns, where the flooring is removed in a sufficiently heated state whereby the intact flooring comes up whole.

Removing asbestos-containing flooring products with an ice scraper is an aggressive removal method and therefore subject to the work practice requirements set forth in Section 7 of this rule; performing this activity requires licensure as an asbestos abatement contractor. Many vinyl sheet floor covering products have a felt-back layering to them; asbestos fibers are present in the layering. The felt is readily visible on the back side of the flooring product. The layers can separate during the removal process thereby releasing asbestos fibers; using infrared heat or heat guns to loosen the product from the substrate may not prevent this (layer separation) from occurring.

(8) Lifting and moving no more than two intact (2' × 4') asbestos-containing suspended ceiling tiles per room to adjacent ceiling tiles and later resetting them back in place to perform routine building maintenance activities above the suspended ceiling system.

(9) Owners collecting samples of asbestos-containing latex, asphaltic, or petroleum-based materials, flooring, and cementitious materials that potentially contain asbestos in their own owner-occupied single-family residence.

Conflict of Interest

Visual evaluations and air clearances for an asbestos abatement project involving more than 100 linear/square feet, or any combination thereof, of ACM must be performed by an Asbestos Consultant firm. The Asbestos Consultant firm must have an independent business relationship with the asbestos abatement contractor or asbestos in-house abatement unit performing the abatement, except as provided in Sections 7(D)(1)(l) and 7(D)(2)(g) of this rule.

A Maine-licensed Asbestos Abatement Contractor may conduct the required visual evaluation and air clearance sampling for an asbestos abatement project under their contractual control that involves less than 100 linear/square feet, or any combination thereof, of ACM provided the individual conducting the visual evaluation and air clearance sampling is a Maine-certified Asbestos Air Monitor.

Notification Requirements

The notification requirements established by this section apply to all asbestos abatement projects except asbestos-associated activities. Demolition activities, excluding single-family residential dwellings, are also subject to the notification provisions of this section. Notification requirements are designed to provide the Department with adequate information to effectively schedule compliance inspections.

▶ GENERAL REQUIREMENTS

The general requirements are as follows.

(1) Unless exempted, any person, owner, or operator engaging in the removal, repair, demolition, enclosure, encapsulation, or handling of more than 3 linear or square feet of an asbestos-containing material must submit written notification of each asbestos project to the Department. The person, owner, or operator is responsible for ensuring that the complete notification, including any applicable fee, is postmarked at least 10 calendar days, or received by the Department at least 5 working days, prior to commencement of the asbestos abatement project, including establishing the regulated area.

Delivery of the notification by U.S. Postal Service, commercial delivery service, hand delivery, or another method as approved by the Department is acceptable. Alternative notification procedures, including those listed in the following, as well as notification of demolitions when no asbestos remains in the building as evidenced by an inspection or complete abatement, may be used.

The start date on the notification should encompass the setup of the regulated area, including any precleaning and the hanging of polyethylene sheeting.

(2) Notification must be on forms approved by the Department, must include all required information under this section, must be accompanied by the appropriate fee, and must be typewritten or easily legible. An incomplete notification is not acceptable and therefore not of record.

(3) The following information must be included in full in the notification:

    (a) A clear indication of whether the notification is the original or a revised notification.

    (b) The name, address, and telephone number of the following:

        (i) The building owner or operator.

        (ii) The asbestos abatement contractor who will perform the asbestos abatement project.

(c) An indication of the type of operation (e.g., demolition, renovation, repair, etc.).

(d) A clear description of the building or affected part of the building including the size (square feet and number of floors), age, and present and prior use of the building.

(e) The procedure, including analytical methods that will be utilized to detect the presence of asbestos material.

(f) The specific amount of ACM to be abated from the building in terms of length of pipe in linear feet or surface area in square feet. All other asbestos-containing materials except for linear lengths of piping must be quantified in square footage.

(g) The building address, including building name, number, and floor or room number of the work area in which the asbestos abatement project will take place, as applicable.

(h) Scheduled starting and ending dates of the asbestos abatement project (encompassing setup, removal, clearance, and tear-down dates).

(i) Scheduled work hours, including planned shift work.

(j) A clear description of demolition, renovation, repair, or other work to be performed and method(s) to be utilized, including specific techniques to be utilized and a clear description of affected components.

(k) A clear description of work practices and engineering controls to be used to comply with the requirements of this rule.

(l) The name and location of the waste disposal site at which the asbestos waste will be disposed.

(m) For a building that is structurally unsound, in danger of collapse, and scheduled to be demolished, the name, title, and authority of the state or local government representative who has ordered the demolition, and the professional engineer who determined that the structure is structurally unsound, the date on which the order was issued, and the date on which the demolition was ordered to begin, along with a copy of that order.

(n) For any other building scheduled for demolition, the dates the demolition is scheduled to occur.

(o) For an emergency asbestos abatement project, the date and hour on which the emergency occurred and a description of the emergency.

(p) The name, address, and telephone number of the transporter scheduled to remove asbestos waste from the site.

(q) A request for approval of any nonstandard work practice(s), including the justification for the request.

(r) Name of the Asbestos Abatement Design Consultant who prepared the original project design for the project.

(s) A contractor or consultant job number consisting of a two- to three-letter company identifier assigned by the Department plus any combination of up to seven letters and digits.

(t) Dates of actual removal/repair activities.

Items (b)(ii), (l), and/or (p) may be noted as unknown on the original notification but must be provided to the Department on a revised notification submitted in accordance with Section 3(D)(2) of this rule. Item (t) may be updated at a minimum 24 hours prior to the new start date for actual removals by telephone contact with Department staff, by fax, or by other methods approved by the Department.

(4) Notification of an asbestos abatement project must be accompanied by a nonrefundable fee paid in full by a cashier's, certified, or company check made payable to the Maine Environmental Protection Fund, or other Department-approved method, in the appropriate amount as follows:

(a) For projects involving more than 100 square feet or 100 linear feet of ACM or any combination thereof, but less than 500 square or 2,500 linear feet of ACM: $100.

(b) For projects involving more than 500 square feet or 2,500 linear feet of ACM, but less than 1,000 square or 5,000 linear feet of ACM: $150.

(c) For projects involving more than 1,000 square feet or 5,000 linear feet of ACM, or any combination thereof of ACM: $300.

(d) For asbestos abatement activities at facilities for which an annual facility notification has been submitted, fees as per (a) and (b) under this paragraph per project. Fees shall be submitted on a quarterly basis.

(e) Fees for condominium units, and individual dwelling units operated as a residential cooperative or military or company housing shall have an annual cap of $5,000. Fees may be submitted on a quarterly basis.

Note: If there are not sufficient funds to cover the check or credit card transaction, an insufficient funds fee will be assessed by the Department in accordance with State of Maine laws and policies. Until that insufficiency is resolved (by money order or bank check), the Department will not accept any additional checks or credit card transactions from the party including additional checks for other project notifications.

(f) Notification fees are not required for asbestos abatement projects occurring in a single-unit residential building.

Alternative Notification Procedures

Alternative notification procedures must be considered in the following cases:

(1) *Emergency Notification.* In the case of an emergency asbestos abatement activity or an ordered demolition of an unsound structure by a state or local authority, the following must be provided:

(a) Oral notification, explaining the event and indicating the need to conduct an asbestos abatement activity, must be made by telephone within 1 working day of the emergency.

(b) Written notification, explaining why the activity qualifies as an emergency and what asbestos abatement activity(ies) will be conducted [including the information required by Section 3(B)(3) of this rule], must be sent by fax, hand delivery, or another method approved by the Department so that it is received by the Department as soon as possible, but no later than 72 hours after the emergency. The fee for the emergency project shall be received no later than 3 days after the emergency notification is submitted.

If the notification fee for an emergency project is not received within 72 hours, the Department will not accept any additional project notifications, license/certification applications, or renewals from the business entity until the emergency notification fee is received by the Department. The Department considers it possible to submit written emergency notification within 24 hours after an emergency in virtually all instances. Department fax machines operate 24 hours per day, 7 days per week.

(2) *Annual Facility Notification.* A facility may notify the Department of asbestos abatement projects on an annual basis. Notification shall be building specific and include the following in addition to the standard notification information required by Section 4(B)(3) of this rule:

    (a) A facility diagram including all buildings in which asbestos abatement projects may take place.

    (b) A description of written method to be used to communicate project dates to the Department at least 24 hours prior to the start of each project.

    (c) A description of the method that ensures that separate standard notification is sent for each abatement project involving more than 160 square feet or 260 linear feet of asbestos-containing material and not forecasted per Section 3(C)(2)(b) of this rule.

    (d) A description of the method or a copy of the form that will be sent to the Department quarterly (calendar year) and that compiles a list of all projects completed and payment of corresponding fees.

    (e) A description of the method(s) that will be used to ensure that nonstandard work practices are not implemented prior to receipt of Department approval.

(3) *Demolition of Buildings.* The demolition of buildings that contain asbestos must be notified as part of the asbestos abatement project notification [see preceding Sections 3.B(3) (m) and (n)]. For the demolition of buildings where it can be demonstrated that no asbestos-containing material is in or on the building, the owner or operator shall notify using alternative forms approved by the Department. Single-family residential buildings are exempt from this notification requirement.

    Intact, nondeteriorated asbestos containing packings, gaskets, resilient floor covering, and asphalt roofing products do not need to be removed from the building prior to demolition if the demolition is performed using large equipment in accordance with the provisions of Section 7(B)(3) of this rule.

(4) *Notification Timeframe Waiver.* Notification for asbestos abatement projects for which the Department approves a notification period less than that required in Section 3(B)(1) of this rule must be received by the Department as soon as possible, but no later than 24 hours prior to commencement

of the asbestos abatement project, including setup or onsite preparation activities. Delivery of the notification by U.S. Postal Service, fax, commercial delivery service, hand delivery, or another method as approved by the Department is acceptable.

To be eligible for this provision, the building owner must demonstrate that reasonable planning and foresight could not have predicted the event and that another notification procedure outlined in these rules would not suffice to protect public health and the environment had it been properly executed. Examples include, but are not limited to, discovering additional asbestos-containing material during a renovation or demolition activity for which a renovation or demolition inspection for asbestos was conducted (e.g., within a wall cavity or plumbing chase), a public health threat exists or will develop if the project is not initiated within a very limited timeframe (e.g., cleanup following a fiber release episode), or conducting a removal project necessitated by an unforeseeable circumstance (e.g., boiler and associated piping/valves failure).

The fee for a Notification Timeframe Waiver project shall be received no later than 3 days after the emergency notification is submitted. If the notification fee for an emergency project is not received within 3 days, the Department will not accept any additional project notifications, license/certification applications, or renewals from the business entity until the emergency notification fee is received by the Department.

Notification Revision Procedures

Notification revision procedures must be adhered to according to the following.

(1) Notification date changes shall be made as follows:
   (a) If the project will begin on a date earlier than the original start date, the owner or operator must submit to the Department a new or revised notification that meets the requirements of Section 3(B)(3) of this rule. This notification must be postmarked at least 10 days prior, or be received by the Department at least 5 working days prior, to the new start date. Delivery by U.S. Postal Service, commercial delivery service, hand delivery, or another method approved by the Department is acceptable.

(b) If the project will begin later, or end earlier or later, than the dates set forth in the original notification, the owner or operator must ensure that the Department receives a new written notification detailing the change(s) in date(s) as soon as possible before, but not later than 24 hours prior to, the original start or actual end date, as applicable. Delivery by U.S. Postal Service, commercial delivery service, fax, hand delivery, or another method approved by the Department is acceptable.

(2) A revised notification form must be sent to the Department for any change(s) to the notification information detailed in Section 3(B)(3) of this rule, including a change of greater than 20% or by more than 100 linear or square feet of the amount(s) of ACM to be affected, whichever is lesser. Delivery may be made by U.S. Postal Service, commercial delivery service, fax, hand delivery, or another method approved by the Department, provided it is received by the Department at least 24 hours prior to the change and prior to completion of the project, except that notification of changes to any nonstandard work practices must be received at least 5 working days prior to implementation of the work practice unless the Department approves a shorter timeframe in accordance with the notification provisions of Section 3(C)(4) of this rule.

Nonstandard work practice may not be implemented without written approval from the Department. Any fee increase for a revised project notification, where applicable, shall be submitted with the notification. Clearances are required for projects exceeding 100 linear or square feet, or any combination thereof, including the change.

▶ LICENSE REQUIREMENTS FOR BUSINESS AND PUBLIC ENTITIES

There are license requirements that need to be addressed, as in the following.

(1) *Scope.* This section sets forth the specific licenses that a business or public entity must obtain prior to engaging in an asbestos abatement activity. This section also sets forth general standards of conduct and specific recordkeeping and other requirements for maintaining each type of license.

(a) A business entity or public entity that engages in an asbestos abatement activity regulated by this rule must hold a valid license as set forth in this section, unless exempted under Section 2(E)(2) of this rule.

(b) A business entity or public entity that engages in an asbestos abatement activity in more than one licensing category set forth in this section must hold a valid license in each category.

(c) An individual engaged in asbestos activities regulated by this rule as a sole proprietor must hold both a valid license and a valid certificate.

(d) A business entity or public entity licensed pursuant to this section must ensure and document that each of its employees is trained in, knowledgeable of, and complies with company-specific standard operating procedures and the requirements of this rule, as applicable.

(e) A business or public entity must maintain all required records at their place of business and must make these available to the Department within 24 hours of request. The business or public entity must also have a written plan for maintaining and archiving records, including provisions for records to be retained for 7 years, even if the licensee ceases business operations.

(2) *Standards of Conduct.* Licensees must comply with all state and federal laws and regulations pertaining to asbestos abatement activities, including the conflict-of-interest provisions of Section 2(G) of this rule. Failure to comply with this rule may result in suspension or revocation of a license, denial of an application for renewal, or other enforcement action deemed appropriate by the Department. Licensees must perform their activities in a manner that

(a) Is in compliance with state-of-the-art professional services generally recognized as acceptable by the asbestos consulting and abatement industries, asbestos professional associations, and government agencies.

(b) Is consistent with current practices taught by Department-licensed Training Providers.

(c) Based on principles, values, standards, or rules of behavior that guide the decisions, procedures, and practices of a licensed entity in a way that contributes to the health and safety of his or her workplace and to all others who may be affected by his or her work.

(3) *General Application Requirements and Procedures*
  (a) An application for a license (including renewal) must be made on forms approved by the Department and must be accompanied by any necessary documentation demonstrating that the licensing requirements of this section have been met.
  (b) An application must be submitted with a nonrefundable application fee paid in full by a cashier's, certified, or company check or other Department-approved payment methods in the amount set forth in this section.
  (c) If an application is incomplete, the Department will either deny it or ask for further information.
  (d) If the Department requests further information from an applicant and does not receive it in full within 30 calendar days, the application will be denied.
  (e) If the Department, after reviewing an application, determines that the applicant has met the applicable requirements of this section, the Department will approve the application and a license will be issued to the business entity or public entity stating the category in which the entity holds a valid license.
  (f) Except as provided at Section 4(A)(4)(c) of this rule, an expired license prohibits the business entity or public entity to which it is issued from engaging in the asbestos abatement activity until a current license is obtained.

(4) *Annual Renewal and Reapplication Procedures*
  (a) A license shall expire 1 year from the date of issuance, except that licenses issued in response to an application submitted within 30 days after the expiration date of a previously issued license will expire 1 year from the expiration date of the previously issued license.
  (b) An applicant may not apply for renewal of a license that has expired more than 30 days.
  (c) If a complete application for renewal of a license is received at least 30 calendar days prior to expiration of the license, the license sought to be renewed will not expire until a final decision has been made by the Department. If a complete application for renewal of a license is not received at least 30 calendar days prior to expiration of the license, the license sought to be renewed will expire until a final decision has been made by the Department.

(d) If an application has been denied under this section, the application may be resubmitted only if the applicant addresses in writing each deficiency given in the denial.

(5) *Denial of Applications.* The Department shall deny an application for a license (including renewal) if the applicant fails to meet the standards established by this rule. Reasons for denial include but are not limited to:

(a) Failure to submit documentation demonstrating its ability to comply with applicable requirements, procedures, and standards set forth in this rule.

(b) Its employees' or agents' history of incompetence or negligence as determined by the Department based on (a) previous compliance inspection(s), review of operating record(s), or other documents.

(c) Submission of false information on an application.

(d) Submission of an incomplete application.

(e) Failure to submit the required fee.

(f) Past violation(s) of state or federal laws or regulations pertaining to asbestos abatement activities or asbestos-associated activities. When issuing a denial, the Department may specify a time period not to exceed 1 year in which the applicant may not reapply for licensure.

(6) *Retention of Records.* Records required by this section shall be maintained for at least 7 years. Records shall be stored at the licensee's normal place of business or an archive or other facility approved by the Department.

(7) *Fees.* License applications must be accompanied by a nonrefundable fee paid in full by a cashier's, certified, or company check or other Department-accepted method, made payable to the *Treasurer, State of Maine*, as follows:

(a) Asbestos Abatement Contractor: $650
   Limited license subcategories include:
   (i) Asbestos-containing exterior cementitious materials
   (ii) Roofing including transite roof shingles
   (iii) Demolition by large equipment

(b) Asbestos Consultant: $650
   Subcategories include:
   (i) Monitoring
   (ii) Inspection
   (iii) Design

(c) Asbestos Analytical Laboratory: $400

(d) In-house Asbestos Abatement Unit: $650
(e) Training Provider: $500 or, with prior written Department approval, the equivalent value of training of Department personnel

If there are not sufficient funds to cover the check or credit card transaction, an insufficient funds fee will be assessed by the Department in accordance with State of Maine laws and policies. Until that insufficiency is resolved (by money order or bank check only), the Department will not accept any additional checks or credit card transactions from the party including checks associated with project notifications.

▶ ASBESTOS ABATEMENT CONTRACTOR LICENSE REQUIREMENTS

A business entity engaged in an asbestos abatement activity must hold a valid Asbestos Abatement Contractor license unless exempted under the provisions of Section 2(E)(2) of this rule. Some activities that may require a valid Asbestos Abatement Contractor license are electric, electronic, plumbing, roofing, siding, flooring, heating, carpentry, masonry, and HVAC activities.

A licensed Asbestos Abatement Contractor engaged in an asbestos abatement activity under its contractual control is not required to hold an Asbestos Consultant license to design, monitor, or collect air samples if performed by an appropriately certified asbestos professional in conjunction with an asbestos abatement activity.

A licensed Asbestos Abatement Contractor must have a certified Asbestos Abatement Project Supervisor employed on staff at all times, except that a Limited Asbestos Abatement Contractor may meet this personnel requirement by subcontracting with an Asbestos Project Supervisor, Asbestos Abatement Design Consultant, or Asbestos Air Monitor services. Employees of licensed Asbestos Abatement Contractors who engage in asbestos abatement or associated activities must be certified pursuant to this rule.

The Department may issue a limited license to an Asbestos Abatement Contractor to engage solely in removal of ACM roofing including transite roof shingles, exterior ACM cementitious materials, or demolition by large equipment with intact ACM flooring.

An applicant for Asbestos Abatement Contractor license including limited licenses must submit information to demonstrate that it meets the requirements of this section and the following:

- A written worker protection program, including a respiratory protection program that conforms with the requirements of OSHA's Respiratory Protection Standard (29 CFR 1910.134 effective April 8, 1998).
- A medical monitoring program that conforms to the requirements of OSHA's Asbestos Standard for Construction (29 CFR 1926.1101, effective August 10, 1994), which includes the identity of the occupational health clinic utilized, number of employees enrolled in the program, and locations of employee exposure records.
- A list of all asbestos-associated citations and notices of violation received in the United States during the last 5 years, including the name of the issuing agency or department, the final disposition of such citation or notice, and, if the applicant's principal owner or operator or officer has received an asbestos-associated citation or notice while owning or operating another company in the previous 5 years, a list of those violations.
- A list of states in which the applicant holds a license, certification, accreditation, or any other approval for asbestos abatement activity.
- A copy of the applicant's standard operating procedures for abatement activities that prevent contamination of a facility and the environment, and that protect the public and employee health from the hazards of exposure to asbestos. Limited licensees must submit standard operating procedures specific to their activities. Roofing firms that intend to remove transite shingles must submit a specific operating procedure for transite shingle removal.
- A copy of the contractor's form for sign-off, by an owner or owner's agent, acknowledging receipt of bulk sampling and project monitoring disclosures is required.
- Proof of access to a licensed asbestos disposal site is required.
- Proof that the applicant's employees engaged in asbestos abatement activities are certified pursuant to the requirements of this rule and a list of the names of the applicant's owner(s), or operator(s), principal(s), and officer(s).

- A list of all other entities performing asbestos abatement activities or asbestos-associated activities of which the applicant, owner or operator(s), principals, or officers are an owner or operator, principal, or officer.
- A list of all names (or acronyms) by which the applicant's firm is known or under which it does business is necessary.
- Any information requested by the Department for purposes of determining the proficiency and adequacy of the applicant's standard operating procedures and proof that at least one employee is a certified Asbestos Abatement Project Supervisor or trained as a competent person for roofing, flooring, exterior cementitious, and demolition by large-equipment projects must be provided.
- A statement affirming that applicable state asbestos rules and regulations, including the recordkeeping requirements of these rules, will be met.

Recordkeeping Requirements

An Asbestos Abatement Contractor must maintain documents set forth under this section at its principal place of business or at an archive facility approved in advance by the Department, in a form that is easily retrievable by project. The documents must be made available to the Department within 24 hours of request.

The name, address, and Department certification number for each of its employees engaged in asbestos abatement activities, including dates of employment, are required. The identification, by name and Department certification number, of each employee's involvement in each of the Asbestos Abatement Contractor's past and present asbestos abatement projects, including name, address, location, and duration of the project, is mandatory. Additional requirements are as follows:

- Copies of all correspondence between the Asbestos Abatement Contractor or its agent and any asbestos regulatory agency including OSHA, for the previous 5 years, including but not limited to letters, notices, citations received, and any notifications made by the contractor pursuant to this rule.
- Copies of all project waste manifests required by the federal NESHAP regulations and Maine's *Non-Hazardous Waste Transporter Licenses,* 06-096 CMR 411.

- Copies of Asbestos Consultants' and Asbestos Analytical Laboratories' reports regarding the project design, onsite project monitoring records, and release of the regulated area including documenting the successful completion of the visual evaluation and air clearance sampling requirements.
- An Asbestos Abatement Contractor must maintain the following items at the abatement work site throughout the duration of such activity and must make the documents immediately available to the Department upon request.
  - A copy of Chapter 425, "Asbestos Management Regulations."
  - A copy of the site-specific asbestos abatement project design.
  - A Department certification card for each onsite employee.
- A daily sign-in log identifying each employee involved in the project by name and Department certification number is required. OSHA requires that you also maintain a daily containment log showing time of entry and egress into the regulated area.
- Records of all onsite monitoring, including personal samples required by 29 CFR 1926.1101, and project documentation.
- A copy of the project notification.
- A copy of the Department approval for any nonstandard work practice granted in accordance with these regulations and a copy of the form signed by the building owner or owner's agent acknowledging receipt of the bulk sampling and/or project monitoring disclosures.

Asbestos Consultant

A business or public entity that engages in the inspection, design, or monitoring of asbestos abatement activities must hold a valid Asbestos Consultant license. Licenses shall specify the function performed. The Asbestos Consultant firm must have an Asbestos Abatement Design Consultant, Asbestos Inspector, or Asbestos Air Monitor on staff at all times, as applicable for the type(s) of services for which they are licensed. Each employee of an Asbestos Consultant who engages in the inspection, project design, or monitoring of asbestos abatement activities must be certified pursuant to these rules.

An applicant for an Asbestos Consultant license must submit sufficient information to demonstrate that the consultant meets the general license requirements set forth in this section and the application requirements of Asbestos Abatement Contractor, except that at least one employee must be certified as an Asbestos Inspector if the firm engages in inspection activities, at least one employee must be certified as an Asbestos Design Consultant if the firm engages in project design activities, and at least one employee must be certified as an Asbestos Air Monitor if the firm engages in air monitoring activities.

Recordkeeping Requirements

An Asbestos Consultant is subject to the contractor recordkeeping requirements 4(B)(5)(a–b) and (i–v), and 4(B)(5)(c)(iii–vi) of this rule to the extent applicable to design, inspection, disclosure, and monitoring, and as further described in Sections 4(D)(4)(b–d) of this rule.

An Asbestos Consultant must maintain copies of daily project logs. Past project logs must be maintained at the principal place of business. Current project logs must be kept up-to-date at the project work site. Project logs include, but are not limited to, sign-in sheets, daily project records, monitoring procedures and data, notifications, work practices associated with the asbestos activity, updated project designs indicating any changes made, and nonstandard work practice(s).

An Asbestos Consultant must maintain the following records at his or her place of business and make them available within 24 hours of request:

- The documents listed in the onsite contractor recordkeeping requirements of these rules as applicable to project monitoring
- Copies of laboratory reports, monitoring documents, and other project documents that may be generated for a particular activity
- Copies of all design documents for each activity
- Copies of standard operating procedures for each activity performed onsite by the Asbestos Consultant
- Copies of any other documents generated in the course of each asbestos abatement activity

An Asbestos Consultant must maintain the following documents at the work site:

- The documents listed in this section
- Current project logs specified earlier

- A copy of the form signed by the building owner or owner's agent acknowledging receipt of bulk sampling and/or project monitoring disclosures

An Asbestos Consultant shall provide an electronic file, or upon request a paper copy, of each 3-year reinspection and management plan recommendation report conducted in accordance with the requirements of AHERA to the Local Education Agency (LEA) and the State of Maine Bureau of General Services (BGS), 77 SHS, Augusta, ME 04333-0077, within 60 days of completion of each reinspection. The Asbestos Consultant shall notify the Department in writing of the reinspection date and the date the report was sent to the LEA and BGS.

▶ ASBESTOS ANALYTICAL LABORATORY

A business or public entity that qualitatively or quantitatively analyzes samples of solids, liquids, or gases for asbestos fibers, or that analyzes air samples for total fiber count, must be licensed as follows:

(a) An Asbestos Analytical Laboratory performing asbestos bulk and/or air analysis of samples collected in the State of Maine must hold a valid license for the type of service provided.

(b) An Asbestos Analytical Laboratory with a license that encompasses air analysis must use phase contrast microscopy (PCM), transmission electron microscopy (TEM), or another EPA-approved method for the analysis of air samples.

(c) An Asbestos Analytical Laboratory with a license that encompasses bulk analysis must use the analytical methods set forth in Section 6(B)(2) of this rule.

(d) An Asbestos Analytical Laboratory that performs air sample analysis must be an active participating laboratory rated proficient by the AIHA's PAT (American Industrial Hygiene Association's Proficiency Analytical Testing) program and must use the analytical methods set forth in Section 8(B)(2)(f) of this rule.

(e) An Asbestos Analytical Laboratory that performs bulk sample analysis must be accredited by the National Voluntary Laboratory Accreditation Program (NVLAP) or be an active participating laboratory rated proficient by AIHA's bulk-quality assurance program. Note: Samples collected as part of a school project must be analyzed by a NVLAP-accredited lab in accordance with federal requirements.

Personnel Requirements

(a) An Asbestos Analytical Laboratory must have on staff at all times a certified Asbestos Air Analyst if the laboratory performs air analyses, or a certified Asbestos Bulk Analyst if the laboratory performs bulk analyses.

(b) Each employee of an Asbestos Analytical Laboratory who engages in work as an analyst must hold a valid Asbestos Air or Bulk Analyst certificate, as applicable.

(c) Each employee of an Asbestos Analytical Laboratory who performs TEM analysis (bulk and/or air) must be properly trained in TEM analytical procedures; documentation of his or her training must be maintained by the laboratory and provided to the department upon request.

Application and Recordkeeping Requirements

(a) Background information on the laboratory, including
   (i) The names of the applicant's owner(s), or operator(s), principal(s), and officer(s)
   (ii) Location and mailing address
   (iii) A list of all other entities performing asbestos abatement activities or asbestos-associated activities in which individuals listed under Section 4(E)(3)(a)(i) of this rule are an owner or operator, principal, or officer
   (iv) A list of all names (or acronyms) by which the applicant's firm is known or under which it does business
   (v) Any information that is requested by the Department for purposes of determining the proficiency and adequacy of the applicant's standard operating procedures

(b) The laboratory must submit documentation describing its QA/QC program for ensuring accuracy of analysis of air and bulk samples. This must include, at a minimum, annual QA/QC training for all Air and Bulk Analysts that includes:

  (i) A review of applicable methods (NIOSH 9002, EPA/600/R-93/116, etc.).

  (ii) A review of relevant current literature (AIHA, ACGIH, McCrone, etc.) and state-of-the-art technology.

  (iii) A review of the lab's current QA/QC program, which should include at a minimum: the statistical calculation of intrinsic sample variability, intracounter variability, and inter- and intralaboratory variability, including the laboratory's current relative standard deviation.

  (iv) A review and hands-on session for microscope cleaning and calibration.

  (v) Two hours of reading QA samples to determine the analyst's proficiency, including actual field samples, round-robin samples from other laboratories if applicable, and third-party QA samples such as the PATs or AARs.

(c) An applicant for an Asbestos Analytical Laboratory must demonstrate that the applicant meets the record-keeping requirements set forth in these regulations and must make the following available for review within 24 hours of request by the Department:

  (i) Sample chain-of-custody procedures, including but not limited to handling, storage, and disposal procedures.

  (ii) A copy of the laboratory's analytical quality assurance program(s).

  (iii) Equipment calibration and standardization procedures.

  (iv) Results of the last four quarters of PAT or round-robin tests, including the following: round number, date of participation in the round, and PAT results including each analyst's individual results.

  (v) Laboratory standard procedures for asbestos analysis.

  (vi) An up-to-date asbestos analytical equipment inventory.

  (vii) Documents related to laboratory personnel training.

  (viii) The certificate of the laboratory owner, operator, or supervisor.

  (ix) A copy of their NVLAP or AIHA accreditation as applicable.

  (x) A copy of their quality assurance program ensuring proficiency of all analysts.

(d) Copies of state certificates and dates of employment for employees performing analyses.

(e) Copies of analyses performed, indicating sample identification number, analysis methods utilized, analytical results, and the name of the certified employee performing the analysis.

▶ IN-HOUSE ASBESTOS ABATEMENT UNIT

A business entity or public entity that engages in asbestos abatement activities solely within the confines of a property owned or leased by that entity solely for its own benefit (not for the purpose of income, profit, or barter) and using its own employees (not independent contractors) must meet the following licensing requirement, as applicable:

- An in-house Asbestos Abatement Unit that engages in asbestos abatement activities, excluding asbestos-associated activities, must meet the application requirements set forth in the "Asbestos Abatement Contractor" section of this rule.
- An in-house Asbestos Abatement Unit that engages in the asbestos-associated activities of inspection, design, and monitoring must meet the application requirements set forth in the "Asbestos Consultant" section of this rule.
- An in-house Asbestos Abatement Unit that engages in the asbestos-associated activity of analysis for asbestos must meet the application requirements set forth in the "Asbestos Analytical Laboratory" section of this rule. In-house laboratories are exempt from laboratory proficiency as long as each analyst is rated proficient by AIHA's Asbestos Analytical Registry or by an active participating laboratory that is rated proficient by AIHA or NVLAP and that has an independent business relationship with the in-house laboratory. Analysts must be annually rated proficient as part of their individual AIHA certification.
- Employees of a licensed in-house Asbestos Abatement Unit who engage in asbestos abatement or associated activities must be certified pursuant to these rules.
- *Recordkeeping requirements.* An in-house Asbestos Abatement Unit must meet the recordkeeping requirements set forth in the "Asbestos Abatement Contractor," "Asbestos Consultant," and "Asbestos Analytical Laboratory" recordkeeping sections of these rules, as applicable.

► TRAINING PROVIDER

A business entity or public entity that provides asbestos training within the geographic boundaries of the State of Maine must hold a valid Training Provider license. A Training Provider whose principal place of business is located outside the State of Maine and who provides training only outside the geographic boundaries of the State of Maine is not subject to the licensing requirements set forth in this section. Training courses also must be approved by the Department under the provisions of Section 10 of this rule.

Personnel Requirements

   (a) *Training Director.* Each Training Provider shall employ a training director, certified as an Asbestos Abatement Design Consultant, who has overall responsibility for all aspects of training.

   (b) *Instructors* must have academic credentials and/or field experience as specified next in the area in which they provide training. Primary instructors must be approved by the Department. If, after receiving a Training Provider license, there is a change in teaching personnel, the Training Provider must notify in writing the Department of the names and credentials of the new instructors at least 30 days prior to the date of the next course offering.

     (i) *Primary Instructors.* The primary instructors are the persons delivering the majority of the training material for the training course and the hands-on portion(s) of the course. Primary instructors must have successfully completed a Department-approved initial training course in the discipline being taught.

     (ii) *Secondary Instructors.* Secondary instructors are persons possessing academic credentials, training, and/or (field) experience in a particular area, who may regularly provide portion(s) of instruction at a course. Secondary instructors do not need to have attended initial asbestos training courses, but need to provide, prior to conducting training, the primary instructor with written documentation detailing their experience and copies of their training and/or academic credentials.

Application Requirements

An applicant for a Training Provider license must submit:

(a) Background information on the Training Provider, including
   (i) The names of the owner(s) or operator(s), principal(s), and officer(s).
   (ii) Location and mailing address.
   (iii) A list of all other entities performing asbestos abatement activities or asbestos-associated activities in which individuals listed in Section 4(F)(3)(a)(i) of this rule are an owner or operator, principal, or officer.
   (iv) A list of all names (or acronyms) by which the applicant's firm is known or under which it does business.
   (v) Any information requested by the Department for purposes of determining the proficiency and adequacy of the applicant's standard operating procedures.
(b) A list of the qualifications and resumes of the instructors, primary and secondary, who will be teaching.
(c) A detailed description of the number and quality of supplies and equipment and the availability of audio/visual teaching aids.
(d) A physical description of the primary training facility, including dimensions, that demonstrates that it is adequate for training and learning purposes as follows:
   (i) Lighting sufficient so that all areas of the training room are adequately illuminated for ease of reading and viewing visual presentations.
   (ii) Room size adequate to accommodate the expected/actual number of attendees with sufficient space to allow attendees ample space for seating, laying out training materials, and exam taking.
   (iii) Tables sized to accommodate each student comfortably, without crowding. Chairs must be comfortable and of proper height to the table(s).
   (iv) Air circulation must maintain a steady exchange of fresh air.
   (v) Background noise must be minimal and not distractive to the learning environment, such that exchanges between the trainer and attendees are audible at all times.

(vi) At a minimum, two means of egress from the building in which the training room is located; a large room may require two means of egress.

(vii) A statement affirming that any facility and/or "hands-on" facility, other than the primary training facility, used for training shall comply with the requirements of this subsection.

(e) A copy of the course sign-in/sign-out log the Training Provider will use to track the times that students arrive and depart the course, including the times out and in for any lunch break. This log is to be filled in by the students when entering and exiting the classroom.

(f) An original student certificate issued upon successful completion of courses.

(g) An example of the format used to communicate to the Department the course results within 5 days of course completion, including participant names, Social Security numbers or date of birth, and exam scores.

(h) A description of the method that the Training Provider will use to notify course dates, times, and location to the Department. All courses must be announced in writing to the Department at least 10 calendar days before the course.

(i) A statement that the Training Provider will issue student certificates within 2 weeks of the completion of the course.

(j) The name and qualifications of the training director.

Recordkeeping Requirements

(a) A Training Provider must maintain records of all requirements of Section 11, course exam(s), and student answer sheets for a period of 7 years and make them available to the Department within 24 hours of request, except as provided in Section 4(F)(4)(b) of this rule.

(b) A Training Provider whose principal place of business is outside the State of Maine and who provides training to individuals seeking a certificate pursuant to this rule must make records and information available to the Department within 5 business days of receipt of a request for information.

Standard of Conduct

The issuance of a fraudulent student certificate, or the violation of any provision of this rule or other applicable laws and regulations including the Model Accreditation Plan (MAP), constitutes grounds for the suspension or revocation of the Training Provider's license, the denial of the renewal of the license, and/or other enforcement action deemed appropriate by the Department.

Reciprocity

Reciprocity, or the acceptance of an individual's training certificate indicating the successful completion of appropriate training, is allowed by this rule. Licensure of Training Providers is not reciprocal.

▶ CERTIFICATION REQUIREMENTS FOR ASBESTOS PROFESSIONALS

This section sets forth specific certification requirements for an individual engaging in an asbestos abatement activity. This section also sets forth the general standards of conduct, specific training, and other requirements for maintaining such certificates.

General Certification Requirements

(1) An individual who engages in an asbestos abatement activity regulated by this rule must hold a valid certification in the discipline appropriate to his or her responsibilities as set forth in this section. The Department is not able to process immediately applications that are hand delivered.

   (a) An individual who engages in asbestos abatement activities in more than one certification category must be certified in each such category.

   (b) A certified individual also must meet all other state, federal, and local accreditation or certification requirements, as applicable.

   (c) An individual must be 18 years of age to be eligible for certification.

(2) Certification by rule. Individuals who are employed by Department-limited licensed companies are considered certified by the Department for purposes of these rules provided such individuals have successfully completed and documented training required by OSHA pursuant to 29 CFR 1926.1101

(effective August 10, 1994) and training documentation is maintained at the abatement work site. Work practices as specified in Section 7 of this rule must be followed by individuals who are certified in accordance with these provisions.

Standards of Conduct

Certified Asbestos Professionals must comply with all state and federal laws and regulations pertaining to asbestos abatement activities, including the conflict-of-interest provisions of Section 2.G of this rule. Failure to comply with this rule may result in suspension or revocation of a certificate, denial of an application for renewal, and/or other enforcement action deemed appropriate by the Department. A certified individual must perform his or her activities in a manner that is

(1) In compliance with state-of-the-art professional services generally recognized as acceptable by the asbestos consulting and abatement industries, asbestos professional associations, and government agencies.
(2) Consistent with current rules and practices taught by Department-approved Training Providers.

Application Requirements and Procedures

(1) *General Application Requirements*
   (a) An application for certification (including renewal) must be made on forms approved by the Department, must include the applicant's Social Security number, and must be accompanied by documentation demonstrating that the substantive certification requirements of this section have been met, including a copy of a training certificate.
   (b) An applicant must submit one passport-size color photo and a nonrefundable application fee paid in full by a cashier's, certified, or company check or other Department-approved payment method, in the amount set forth in this section.
   (c) An applicant for certification must submit sufficient documentation to demonstrate that he or she has successfully completed the discipline-specific training and experience requirements. Primary instructors must attend an initial training course from a licensed Training Provider who has an independent business relationship with the primary

instructor, but may attend refresher courses taught by any licensed Training Provider. The primary instructor must notify (on the notification required by these regulations) the Department when he or she will attend his or her own refresher course. Primary instructors are prohibited from conducting refresher courses for themselves only.

It is the responsibility of the individual planning to attend a training course to ensure that the training course is approved by or acceptable to the Department. A training course that is not approved by the Department will not meet the standards for certification of individuals pursuant to this section unless reciprocity is granted.

(d) An applicant must attest knowledge of and compliance with his or her current employer's standard operating procedures.

(e) The Department will either deny an application or ask for further information.

(f) If the Department requests further information and does not receive it in full within 30 calendar days, the application will be denied.

(g) After reviewing an application and determining that the applicant has met the minimum requirements of training and experience (where applicable) as set forth in this section, the Department will approve the application and issue a state certificate.

(h) A state-issued asbestos certification card evidencing that the individual is currently certified to perform asbestos abatement activities is the property of the individual to whom it is issued.

(i) Except as provided in this section, an expired certificate prohibits the individual from engaging in the applicable asbestos abatement activity until a current certificate is obtained.

(2) *Annual Renewal and Reapplication Procedures*

(a) A certificate shall expire 1 year from the last day of the month from the date of issuance, or on the last day of the month that the training certificate expires, whichever is sooner, and may be renewed on an annual basis.

(b) If the Department receives a request at least 5 working days prior to a certification expiration date, the Department may grant an extension of a certificate for up to 30 days when a Maine refresher course has not been available within the last 30 days. Extensions to certifications are not valid for work on projects in schools and public and commercial buildings.

(c) If a complete renewal application is received by the Department at least 5 working days prior to expiration of their state certificate, the certificate will not expire until the Department takes final action on the application.

(d) Individuals must take an annual refresher course or participate in a laboratory quality assurance program as appropriate to their respective certification category. Individuals who fail to meet this requirement within 1 year of the expiration date of their training certificate must take initial training again.

(e) If an applicant has been denied under this section, the applicant must adequately address in writing each denial reason.

Denial of Applications

The Department may deny an applicant for the following:

(a) Failure to submit documentation demonstrating his or her ability to comply fully with applicable requirements, procedures, and standards set forth in this rule.

(b) A history of incompetence or negligence, as determined by the Department based on previous compliance inspection(s), review of operating record(s), or other documents.

(c) Submission of false information on an application.

(d) Submission of an incomplete application.

(e) Failure to submit the required fee.

(f) Past violations(s) of state or federal laws or regulations pertaining to asbestos abatement activities.

Reciprocity

An individual who is certified, accredited, or permitted by another governmental agency may be granted reciprocity from the Department for a certificate in the State of Maine. The applicant must submit a nonrefundable application fee, as set forth in this section, along with appropriate documentation. The certificate will be granted only if the Department determines the certification requirements of the other governmental agency to be at least as stringent as the applicable requirements of this rule, including approval of the training course by an EPA-approved state program or equivalent pursuant to Appendix C, the Interim Final Rule-59 FR 5236-5260 (effective April 4, 1994), to Subpart E of the Asbestos-Containing Materials in Schools rule, 40 CFR, Part 763 (effective December 14, 1987).

Fees

Applications for initial and renewal certification must be accompanied by a nonrefundable application fee paid in full by a cashier's, certified, or company check or other Department-accepted method, made payable to the *Treasurer, State of Maine*, as follows:

(a) Application fees:
  - Asbestos Abatement Worker: $50
  - Asbestos Abatement Project Supervisor: $100
  - Asbestos Air Monitor: $100
  - Asbestos Inspector: $100
  - Asbestos Abatement Design Consultant: $100 (includes limited certification)
  - Asbestos Air Analyst: $100
  - Asbestos Bulk Analyst: $100
  - Asbestos Management Planner: $100
(b) An individual applying for a certificate in more than one certification category during the same calendar year must pay the fee for the highest category first and then pay $50 for each additional category.
(c) Reissuance of a certificate or photo ID card: $50

If there are not sufficient funds to cover the check or credit card transaction, an insufficient funds fee will be assessed by the Department in accordance with State of Maine laws and policies. Until that insufficiency is resolved (by money order or bank check), the Department will not accept any additional checks or credit card transactions from the party including checks for project notifications.

Required Certification

An individual who engages in asbestos abatement activities, excluding associated activities, must hold either a valid Maine Asbestos Abatement Worker or a Maine Asbestos Abatement Project Supervisor certificate. Individuals who directly supervise those engaged in any asbestos abatement activity, excluding associated activities, must hold a valid Maine Asbestos Abatement Project Supervisor certificate.

Any individual who documents and/or oversees an asbestos abatement activity for an owner or operator, conducts visual evaluations of the work area, collects air samples, or conducts project monitoring must hold a valid Maine Asbestos Air Monitor certificate. People who engage in inspections or survey facilities to identify and/or assess the condition of ACM, or an individual who collects bulk samples for analysis, must hold a valid Maine Asbestos Inspector certificate.

An individual who engages in any one or more of the following asbestos-associated activities must hold a valid Maine Asbestos Abatement Design Consultant certificate:

(a) Develops project designs for asbestos abatement or facility renovation, repair, or replacement where the work will impact or has the potential to impact or disturb ACM regulated by this rule.

(b) Supervises the implementation of project designs.

(c) Designs and supervises training courses as a Training Director.

Individuals who perform analyses of air samples for asbestos or total fiber count must hold a valid Maine Asbestos Air Analyst certificate. A worker who performs analyses of bulk samples for the quantification or qualification of asbestos must hold a valid Maine Asbestos Bulk Analyst certificate. An individual who uses data gathered by an Asbestos Inspector to assess asbestos hazards, determine appropriate response actions, or develop asbestos response action implementation or asbestos operation and maintenance programs in schools must hold a valid Maine Asbestos Management Planner certificate.

Training, Education, and Experience Requirements

The minimum training, education, and experience requirements for initial certification in each discipline are as follows:

(a) *Asbestos Abatement Worker*—Successful completion of a Department-approved 4-day (32-hour) asbestos abatement worker training course and exam.

(b) *Asbestos Abatement Project Supervisor*—Successful completion of a Department-approved 5-day (40-hour) asbestos abatement project supervisor training course and exam.

(c) *Asbestos Air Monitor*—Successful completion of a Department-approved 5-day (40-hour) asbestos air monitor training course and exam; or successful completion of Department-approved project supervisor and 16-hour (minimum) air monitoring courses and exams; or, for out-of-state applicants, a combination of training and experience equivalent to the training required in this section.

(d) *Asbestos Inspector*—Successful completion of a Department-approved 3-day (24-hour) inspector training course and exam.

(e) *Asbestos Abatement Design Consultant*
  (i) Successful completion of a Department-approved 3-day (24-hour) project design training course and exam.
  (ii) Possession of at least a bachelor's degree, or possession of any valid asbestos professional certificate, excluding Asbestos Abatement Worker, for 3 years, or some other combination of training, education, and experience deemed appropriate by the Department.

(f) *Asbestos Abatement Design Consultant*—Limited Certification for Training Directors
  (i) Successful completion of a Department-approved 3-day (24-hour) project design training course and exam.
  (ii) A postsecondary degree in adult education or successful completion of a "Train the Trainer" course acceptable to the Department consisting of the following course topics at a minimum:
    • Adult learning principles
    • Training theory
    • Various training skills and techniques
    • Delivery techniques
    • Organizing and preparing presentations
    • Classroom management
    • Small and large group activities
    • Use of audio visuals

(iii) Possession of at least a bachelor's degree, or possession of any valid asbestos professional certificate, excluding Asbestos Abatement Worker, for 3 years, or some other combination of training, education, and experience deemed appropriate by the Department.

Asbestos Air Analyst must successfully complete a Department-approved 5-day (40-hour) Asbestos Air Analyst training course and exam equivalent to the former NIOSH Course #582, "Sampling and Evaluation of Airborne Asbestos."

Asbestos Bulk Analyst must complete a Department-approved training course and exam successfully in the techniques and procedures for identification and quantification of asbestos in bulk samples (e.g., McCrone Institute Asbestos Bulk Analysis course or its equivalent).

► ASBESTOS MANAGEMENT PLANNER

All asbestos management planners must comply with the following:

(i) Successful completion of a Department-approved 3-day (24-hour) inspector training course and exam.

(ii) Successful completion of a Department-approved 2-day (16-hour) management planner training course and exam.

Certification Renewal

The minimum ongoing training and other requirements for renewal certification in each discipline are as follows:

(a) *Asbestos Abatement Worker*—Successful completion of a Department-approved 1-day (8-hour) worker annual refresher course and exam.

(b) *Asbestos Abatement Project Supervisor*—Successful completion of a Department-approved 1-day (8-hour) project supervisor annual refresher course and exam.

(c) *Asbestos Air Monitor*—Successful completion of a Department-approved half-day (4-hour) asbestos air monitor refresher course and exam.

(d) *Asbestos Inspector*—Successful completion of a Department-approved half-day (4-hour) asbestos inspector refresher course and exam.

(e) *Asbestos Abatement Design Consultant*—Successful completion of a Department-approved 1-day (8-hour) project design refresher course and exam.

(f) *Asbestos Air Analyst*
    (i) Documentation of participation in a Maine-licensed laboratory's annual analyst QA/QC training.
    (ii) A statement from a licensed Asbestos Analytical Laboratory affirming that the applicant is proficient in air analysis through participation in the laboratory's administration of the AIHA's PAT program, including results of the last four rounds of the applicant's testing, or proof that the applicant is listed in the AIHA Asbestos Analysis Registry.

(g) *Asbestos Bulk Analyst*
    (i) Documentation of participation in a Maine-licensed laboratory's annual analyst QA/QC training.
    (ii) A statement from a licensed Asbestos Analytical Laboratory affirming that the applicant is proficient in bulk analysis through participation in the laboratory's administration of the NVLAP or AIHA bulk quality assurance program, including results of the last four rounds of the applicant's testing.

(h) *Asbestos Management Planner*—Successful completion of a Department-approved half-day (4-hour) asbestos manager planner refresher course and exam.

Responsibilities

The responsibilities of Maine asbestos workers are as follows:

(1) An *Asbestos Abatement Worker* must possess a valid Maine certificate at the work site.

(2) An *Asbestos Abatement Project Supervisor* must possess a valid Maine certificate at the work site. An Asbestos Abatement Project Supervisor who is the lead supervisor on a project is responsible for directing correction of problems when encountered and terminating activity if needed to comply with all applicable work practice requirements of these rules.

(3) An *Asbestos Air Monitor*
    (a) Must conduct monitoring activities in accordance with recognized monitoring practices. This includes but is not limited to:

      (i) Clearance sampling techniques, including those required by TEM

     (ii) Location and frequency of area monitoring including background sampling

    (iii) Pump calibration methods

    (iv) Settled dust sampling techniques

     (v) Visual evaluation techniques

    (vi) Project monitoring including preparation of an air monitor plan as part of a project design

  (b) May interpret project specifications or abatement plans and monitor and evaluate asbestos abatement activities for compliance with applicable laws and regulations.

  (c) Must possess a valid Maine certificate onsite at an asbestos abatement activity.

(4) An *Asbestos Inspector*

  (a) May review building records and perform inspections of facilities to identify, document, or inventory materials suspected of containing asbestos. An Asbestos Inspector shall collect bulk samples for asbestos analysis according to generally recognized procedures established by the industry, the Department, current U.S. EPA guidance documents, and other applicable state or federal rules or regulations.

  (b) Must, when performing assessments or evaluations of ACM, utilize assessment methodologies outlined in AHERA or an equivalent assessment system to evaluate the condition, accessibility, and intactness of ACM.

  (c) Must possess a valid Maine certificate onsite at an asbestos abatement or associated activity.

(5) An *Asbestos Abatement Design Consultant*

  (a) Must apply knowledge of building construction, design, and use to develop operations and maintenance plans, abatement designs including air monitoring plans, specifications, bidding documents, architectural drawings, and schematic representations of material locations, in a manner consistent with these regulations.

  (b) Must possess a valid Maine certificate onsite at an asbestos abatement activity or training course location.

(6) An *Asbestos Air Analyst* must possess a valid Maine certificate at any facility or location where analysis is being performed.

(7) An *Asbestos Bulk Analyst* must possess a valid Maine certificate at any facility or location where analysis is being performed.

(8) An *Asbestos Management Planner* must possess a valid Maine certificate at his or her principal place of business.

▶ PRE-ABATEMENT REQUIREMENTS

Prior to conducting a renovation or demolition activity that impacts any building material likely to contain asbestos (e.g., those used in roofing, flooring, siding, ceiling, and wall systems) or any component likely to contain asbestos (e.g., heating, ventilation, air conditioning, and plumbing systems), the owner or operator must have an inspection conducted for the presence of asbestos-containing materials. In lieu of inspection, the owner or operator may presume that building materials and components contain asbestos that requires these materials be abated in accordance with these rules.

A DEP-certified Asbestos Inspector must perform the inspection. The inspection must identify all asbestos-containing materials that could be impacted during the renovation or demolition activity, must be completed prior to submission of notification to the Department, must be in writing, and must be onsite and made immediately available to the Department upon request.

Residential dwellings constructed before 1981 that consist of two (2) to four (4) units must be evaluated for building materials and components that are likely to contain asbestos. This evaluation may be performed by a DEP-certified Asbestos Inspector or by a person familiar with asbestos-containing building materials. If building materials and/or components likely to contain asbestos are found, these must be removed in accordance with these regulations prior to demolition except as allowed in Section 7(B) of this rule or must be tested by a DEP-certified Asbestos Inspector to demonstrate that they are not ACM.

Single-family residences and those residences constructed after 1980 that consist of two (2) to four (4) units, are exempt from the inspection provisions of this section. Specific building materials that do not require inspection, sampling, and analysis for asbestos include wood, fiberglass, glass, plastic, metal, laminates, foam, rubber, and gypsum board when joint compound was used only as a filler and not as a layered component, and intact caulkings and glazings. Also, building materials do not need to be inspected when written documents exist confirming that no asbestos was used in the materials that will be impacted, or that the materials were previously inspected by a DEP-certified Asbestos Inspector and affirmatively determined through sampling and analysis to not be ACM.

To maintain compliance with Maine law, if more than 3 square feet or 3 linear feet of ACM are present, this ACM must be removed prior to the demolition, except that intact packing, gaskets, roofing, and flooring may be left in place when the demolition is performed by large equipment in accordance with these rules. Homeowners are encouraged to conduct a walkthrough of their single-family homes to identify suspect asbestos-containing materials, such as thermal system insulation, ceiling tile, exterior cementitious siding, rigid panels, and flooring, and hire an Asbestos Consultant or Asbestos Abatement Contractor if suspect materials are observed. The Department can provide, upon request, more information regarding common asbestos-containing materials in buildings.

Inspection Requirements

Inspection includes collecting bulk samples for analysis and/ or conducting assessments of ACM. Inspections must be conducted as follows:

(1) Bulk samples must be collected by a Department-certified Inspector as prescribed in the following, in a random manner such that they are representative of each homogenous area. Bulk samples shall be collected and analyzed for all asbestos abatement activities unless an approved disclosure is received by the owner or owner's agent from the operator prior to the start of the project.

(a) From Surfacing Material:

    (i) 3 bulk samples from each homogenous area and/or material that is 1,000 square feet or less

    (ii) 5 bulk samples from each homogenous area that is greater than 1,000 square feet but less than or equal to 5000 square feet

    (iii) 7 bulk samples from each homogenous area that is greater than 5,000 square feet

(b) From Thermal System Insulation:

    (i) 3 bulk samples from each homogenous area

    (ii) 1 bulk sample from each homogenous area of patched thermal system insulation if the patched section is less than 6 linear or square feet

(iii) Samples sufficient to determine whether the material is ACM from each insulated mechanical system where cement is utilized on tees, elbows, or valves

(c) From Miscellaneous ACM:

   (i) 3 samples from each miscellaneous material

   (ii) 1 sample if the amount of miscellaneous material is less than 6 square or linear feet

(d) An Asbestos Consultant may implement an alternative sampling protocol that collects more but not less than the number of samples per homogeneous area set forth in this section, provided the Asbestos Consultant has informed the building owner or owner's agent in writing of the standard sampling protocol set forth prior to the sampling event. The Asbestos Consultant must document that the building owner or owner's agent received information regarding the standard sampling protocol set forth in this section by obtaining the building owner's or owner's agent's signature on a statement acknowledging receipt of the information before the sampling event begins. The Department will provide Asbestos Consultants with preapproved disclosure language regarding the standard sampling protocol.

(2) Analysis. Bulk samples collected pursuant to this rule must be analyzed by a Department-licensed Asbestos Analytical Laboratory as described next.

(a) Bulk samples shall be analyzed until a positive result is obtained or all samples have been analyzed. Reanalysis is not required if the sample result is less than 1%.

(b) Wherever there is a suspect asbestos-containing material and a mastic/adhesive affixed to that material, the mastic/adhesive shall be analyzed and reported separate from the suspect asbestos-containing material.

(c) Analysis of suspect asbestos-containing surfacing materials, thermal system insulation, and cementitious materials.

   (i) Bulk samples of surfacing materials and thermal system insulation and cementitious materials shall be analyzed using the PLM–EPA 600/R-93/116 visual estimation method (1993).

   (ii) Point counting surfacing materials and thermal system insulation samples. The Asbestos Consultant shall advise the building owner or owner's agent whenever the Asbestos Analytical Laboratory has reported friable bulk samples with an asbestos content of less than 10% using the standard visual estimation PLM–EPA 600/R-93/116 method. The building owner or the owner's agent may either elect to treat the bulk material as asbestos-containing with no point counting required, or may request that the laboratory further characterize asbestos percentage by using a point count method. Point counting methods are as follows:

- PLM EPA/600/R-93/116 (200 Point Count)
- PLM EPA/600/R-93/116 (400 Point Count)
- PLM EPA/600/R-93/116 (1,000 Point Count)

(iii) Alternative analytical method for suspect asbestos-containing surfacing materials, thermal system insulation, and cementitious materials. The Asbestos Consultant shall advise the building owner or the building owner's agent whenever the Asbestos Analytical Laboratory has determined that is it not feasible or appropriate to have bulk sample(s) of suspect asbestos-containing surfacing materials, thermal system insulation, and cementitious materials analyzed using the standard visual estimation PLM-EPA 600/R-93/116 method: The building owner or the building owner's agent may then either elect to treat the suspect bulk material(s) as asbestos-containing with no further analysis required, or may consent to the use of an alternative analytical method, EPA 600/R-93/116 Section 2.5.5.2 (TEM % by Mass), to determine whether the suspect bulk sample(s) is asbestos-containing.

(d) Analysis of asbestos-containing nonfriable organically bound materials (NOB) and bulk samples of NOB, including but not limited to floor tiles, asphalt shingles, caulking, glazing, mastics, coatings, sealants, adhesives, and glues shall be analyzed using PLM NOB–EPA 600/R-93/116 with gravimetric preparation method.

(i) Point counting NOB samples:
The Asbestos Consultant shall advise the building owner whenever the Asbestos Analytical Laboratory has reported an NOB sample with an asbestos

content of less than 10% using PLM NOB–EPA 600/R-93/116 with gravimetric preparation method. The building owner may either elect to treat the NOB sample as asbestos-containing with no point counting required, or may request that the laboratory further characterize asbestos percentage by using a point count method. The analyst shall point count the sample residue after the gravimetric preparation is completed and/or in accordance with the analytical method.

(ii) Alternative analytical methods for nonfriable organically bound materials (NOB) samples:

The Asbestos Consultant shall advise the building owner whenever the Asbestos Analytical Laboratory has determined it is not feasible or appropriate to have suspect bulk samples of NOBs analyzed using the standard PLM–EPA 600/R-93/116 with gravimetric preparation. The building owner may then either elect to treat the suspect bulk material(s) as asbestos-containing ones with no further analysis required, or may consent to the use of an alternative analytical method to determine whether the suspect bulk sample(s) is asbestos-containing. There are some alternative methods available.

An Analytical Laboratory may use TEM, or another Department-approved analytical method, for bulk sample rather than the standard PLM analytical method set forth in this section. Asbestos Consultants who collected the bulk samples for the building owner must document that the building owner or owner's agent received information regarding the standard analytical protocol set forth in this section by obtaining the building owner's or owner's agent's signature on a statement acknowledging receipt of the information before the sample analysis for TEM analysis begins. The Department will provide Asbestos Consultants with preapproved language regarding standard analytical protocol disclosure.

In instances where there is a positive and a negative sample result for the same sampled bulk material(s) from different sampling events, the material(s) is considered to be asbestos-containing. If a building owner elects to resolve the discrepancy between the two analytical results by resampling of the material, the building owner must either have both Asbestos Inspectors from the previous sampling events present at the resampling of the material(s), or the building owner may elect to have a third party resample the material(s). The building owner must provide the third party with the inspection reports and analytical results from the earlier sampling events before resampling. In either instance, split samples shall be sent to two separate laboratories for reanalysis using an agreed-on analytical method; any sample/material testing positive by either laboratory is positive for asbestos.

Sampling and analysis of nonregulated materials and media including but not limited to water, dust, rock, soil, minerals, and asbestos-contaminated products such as vermiculite may be appropriate to determine the presence of asbestos fibers in the material or medium and to assist in determining appropriate work practices and the scope of any clean-up activities of these nonregulated materials and media. Current state-of-the-art analytical methods include:

(a) Analysis of water samples for asbestos—that is, EPA method 100.2 analytical method
(b) Analysis of dust samples for asbestos
(c) Analysis of rock, soil, and mineral such as vermiculite samples for asbestos:
  • CARB 435 Level A-C (preferred method)
  • EPA 600/R-93/116
  • EPA-600/M4-82-020; (40 CFR, Appendix A to Subpart E)
  • Region I SOP:EIA-INGASED3.SOP (3/9/05)
  • EPA 600/R-04/044 (PLM and TEM)

There are instances where there is visible bulk asbestos-containing materials that get comingled with soil. For example, asbestos-containing thermal system insulation in crawl spaces will fall off the piping onto the bare soil below. Similarly, there are instances where other asbestos-containing materials become comingled in soils such as asbestos-containing bulk materials that are buried or disposed of in a manner or location that is in violation of Department or other applicable regulatory standards such as NESHAP; the removal of greater than 3 square feet of visible asbestos-containing bulk material comingled in soils is an activity that is regulated under this rule.

Design Requirements

Asbestos abatement activities, exclusive of asbestos-associated activities, are subject to the project design requirements specified in the following. The Asbestos Abatement Design Consultant responsible for the design of record (latest version) must ensure that the design is consistent with the requirements of this rule. The design must be completed prior to the start of the activity. Any change to the original project design must be approved in writing by an Asbestos Abatement Design Consultant. The project design must contain the following at a minimum:

(1) Project drawings, including ACM location and quantity, location of regulated area, ventilation system, decontamination facility, and, where applicable, the waste load-out unit.
(2) An air monitoring plan developed in accordance with the requirements of Section 8, and including the number of air clearance samples included in the project contract.
(3) A description of work practice methods to be used.
(4) A description of personnel decontamination methods and sequences.
(5) Documentation of any nonstandard work practices including documentation of the notification to the Asbestos Abatement Design Consultant who prepared the original project design for the project of any nonstandard work practice requests submitted to the Department.
(6) If bulk sampling and/or project monitoring will not be performed, copies of the appropriate disclosures, as developed by the Department.
(7) The name, certification number, and signature of the Asbestos Abatement Design Consultant responsible for the design.
(8) The function(s) and respective areas of responsibility for certified personnel and licensed companies involved in the project.
(9) A description and scope of work for precleaning any existing asbestos-containing debris within the boundaries of the regulated area, as defined by the design.

Disclosures

Prior to agreeing to perform an asbestos-associated activity, an Asbestos Consultant firm must provide to the facility owner or agent a disclosure form approved by the Department listing all asbestos abatement contractor firms with which the Asbestos Consultant does not have an independent business relationship.

Prior to the start of the asbestos abatement activity, if any materials are presumed to contain asbestos, the Asbestos Abatement Contractor or the Asbestos Abatement Design Consultant (as determined by the design requirements) must provide the building owner or building owner's agent with a bulk sampling disclosure listing which materials identified for abatement have not been sampled and the cost of sampling those materials.

Also prior to the start of the asbestos abatement activity, the asbestos abatement contractor or the Asbestos Abatement Design Consultant (as determined by the design requirements) must provide the building owner or building owner's agent with a project monitoring disclosure form discussing the advantages of project monitoring, including project management and/or area sampling, by an Asbestos Consultant firm with an independent business relationship with the entity performing the abatement.

The bulk sampling disclosure and the project monitoring disclosure forms must be approved by the Department before they may be provided to the building owner or building owner's agent. Operators must document that the building owner or owner's agent received the disclosure(s) by obtaining the building owner's or owner's agent's signature on a statement acknowledging receipt.

Disclosures may be submitted to the owner or owner's agent on an annual basis for facilities with ongoing asbestos abatement activities. The Department will provide Asbestos Abatement Contractors and Asbestos Abatement Design Consultants with preapproved language for the bulk sampling disclosure and the project monitoring disclosure.

▶ ASBESTOS ABATEMENT WORK PRACTICE REQUIREMENTS

This section establishes minimum work practice requirements for asbestos abatement activities. All asbestos abatement activities are subject to these work practice requirements. All projects must be performed in accordance with an applicable project design, as set forth in these rules.

General Work Practice Requirements

Asbestos abatement activities, exclusive of asbestos-associated activities, must comply with the following work practice requirements:

(1) A *Certified Asbestos Abatement Project Supervisor* must be designated as the lead supervisor for the project and must be present at the work site at all times personnel are within the regulated area.

(2) *Establishing the Regulated Area.* Prior to starting an asbestos abatement activity, the Asbestos Abatement Contractor must establish the regulated area. For activities where containment is not required, the regulated area shall be demarcated with barrier tape marked "ASBESTOS HAZARD" (or equivalent wording) and OSHA warning signs, and located such that it protects persons from exposure to asbestos and minimizes the number of persons in the area. In facilities where plastic barrier tape may cause a safety hazard, red cloth tape may be used.

If establishing an exclusion zone, the exclusion zone should be demarcated with barrier tape that is different from the tape used to demarcate the regulated area. The regulated area must include the following:

(a) Except as allowed under the provisions of sections 7(B) and 7(D) of this rule, a work area containment meeting the following requirements:

(i) A polyethylene-enclosed structure formed by partitions or framing or by covering walls and ceilings with a minimum of two layers of 4-mil polyethylene sheeting or one layer of 6-mil polyethylene sheeting, and by covering the floor with a minimum of two layers of 6-mil polyethylene sheeting. The surface to be abated does not need to be covered with polyethylene sheeting. For suspended ceiling tile system removals, containment above the ceiling also is required for interior walls that do not extend from floor to ceiling; that is, a gap exists above the ceiling system where asbestos fibers could migrate to areas not being abated. Perimeter areas along interior walls therefore must be accessed and/or removed first and two layers of 4-mil polyethylene sheeting or one layer of 6-mil polyethylene sheeting established as containment prior to removing interior portions of the ceiling system. Exterior walls must have critical barriers established in Section 7(A)(2)(d) of this rule.

(ii) Fiber-tight seams in the polyethylene coverings.

(iii) An access into the polyethylene-enclosed containment provided through the decontamination unit.

   When containment is not required, all other work practice requirements, including all other requirements for the regulated area, still apply.

(b) A decontamination facility consisting of aluminum, tin, fiberglass, preformed plastic, or other impervious surface, or two layers of 6-mil polyethylene sheeting. Decontamination facilities must have 6-mil polyethylene sheeting flaps or air-locks between each chamber. Remote decontamination facilities are exempt from the ventilation system required below. Where construction of a decontamination unit meeting minimum size requirements is not possible due to room size and configuration, HVAC system component locations, life safety code requirements, or restriction of safe egress for residents, a smaller than standard decontamination facility may be constructed.

(c) A ventilation system meeting the following requirements:

   (i) The exchange of at least 4 volumes of air per hour at a volume sufficient to establish and maintain a pressure differential within the ambient environment of negative 0.02 inches of water column.

   (ii) The ventilation units must be operated in accordance with U.S. EPA recommendations set forth in Appendix J of U.S. EPA Guidance Document EPA 560/5-85-024 (effective June 1985) or in Appendix F to 29 CFR, Part 1926.1101 (effective August 10, 1994).

   (iii) The make-up air entering the containment must pass through the decontamination system whenever that is possible, or through waste load-out and/or make-up air intakes specified by the project design.

   (iv) The exhaust air must be HEPA filtered before being discharged outside of the work area and must be discharged outside the facility to a location that is not near any intake for building ventilation. The HEPA-filtered exhaust air may be discharged inside the facility if access to the outside is not feasible because the distance from the regulated area to the outside of the building is too great (such as in large industrial buildings or warehouses), or when health and safety concerns (such as blocking egress from an area with limited access).

If the exhaust air is discharged inside the facility, the contractor shall demonstrate that the unit(s) is operating effectively by evidencing that air samples collected from the exhaust stream are less than 0.01 f/cc or that the audio alarm filter and/or filter change lamps and the unit(s) pressure differential filter monitoring meter are operational.

Ventilation units may be shut down overnight when the only source of electrical power for the project is a portable generator that must be removed at night for site safety and security reasons. The entry into the work area must be sealed (fiber-tight) whenever ventilation units are shut down.

(v) The exhaust air tubes or ducts associated with the work area ventilation system must be fiber-tight and must be securely attached to the HEPA unit exhaust port.

(d) Critical barriers are required for all projects.

(i) Prior to suspended ceiling tile removals, the perimeter area above the ceiling must be accessed first, under negative pressure with properly protected employees to allow the sealing, as critical barriers, of penetrations and openings along the perimeter. Removal of perimeter ceiling tiles must be conducted as an abatement activity subject to all applicable work practice requirements.

(e) A waste load-out, if applicable.

(3) *Exclusion of Persons from the Regulated Area and Posting Signs*

(a) Individuals not directly involved in the asbestos abatement activity must be excluded from the regulated area.

    (b) Warning signs, meeting the requirements set forth in 29 CFR 1926.1101 (effective August 10, 1994), shall be posted at all approaches to the regulated area, and at the decontamination and waste load-out unit's outermost boundaries.

(4) *Regulated Area Control.* An Asbestos Abatement Project Supervisor must be present at the work site at all times during active abatement activities. The Supervisor must have the authority to initiate and implement corrective action should problems or deficiencies arise at the asbestos work site.

(5) *Physical Barriers.* Physical barriers must be established if indicated in the project design.

(6) *Personal Protective Equipment.* An individual involved in an asbestos abatement activity or an individual who enters the regulated area, excluding the clean room, must be provided with and wear appropriate respiratory protection and personal protective clothing. Minimum respiratory protection shall be a half-face negative-pressure respirator equipped with HEPA filters. Minimum protective clothing shall be disposable full-body suits, including head and foot coverings.

    Wearing a nylon brief–type bathing suit underneath disposable full-body suits is allowed. Gym shorts, "cut-offs," or underwear are not considered to be bathing suits. OSHA also regulates asbestos activities involving respirators and personal protective equipment. OSHA regulations may require a higher degree of respiratory protection and/or protective clothing.

(7) *Isolation of HVAC Systems.* All intake openings, exhaust openings, and any holes in the building HVAC system and its components located within the regulated area must be fiber-tight and covered with two layers of 6-mil fiber-tight sheeting, and all seams in the system must be taped to be fiber-tight.

(8) *Covering of Movable and Immovable Objects*

    (a) Movable objects within the regulated area must be removed or, if not feasible, treated as an immovable object below.

    (b) Immovable or fixed objects within the regulated area must be wrapped with two layers of 6-mil (minimum) polyethylene sheeting that is fiber-tight prior to the commencement of abatement activity.

(9) *Air and Project Monitoring.* All asbestos abatement activities are subject to the following air and project monitoring requirements:

   (a) An air monitoring program that is developed by an Asbestos Air Monitor must be in place and be consistent with these rules.

   (b) A project monitoring program, if applicable, that is developed by an Asbestos Air Monitor must be in place and be consistent with these rules.

      An air monitoring program consists of air clearance sampling at a minimum, and may include background and area samples. OSHA requires personal sampling for most activities.

(10) *Wetting of ACM.* Prior to removal of ACM, including removal of components covered with thermal system insulation, all ACM must be adequately wetted with water, except as provided next. Throughout the removal, storage, transport, and disposal processes, ACM must be kept adequately wet.

   (a) Wetting ACM is not required under the following conditions:

      (i) When the temperature inside the regulated area is below 32°F and heating the area is neither feasible nor practical (e.g., abandoned warehouse or roof).

      (ii) When electrical conditions exist that are noted in the design plan and demarcated in the specific work area, and that would render that specified area hazardous to shock and/or electrocution hazards.

      (iii) When operational high-pressure steam lines are being abated or repaired.

      (iv) Wetting metal jacketed piping during wrap-and-cut operations. ACM exposed during glove bagging associated with the wrap-and-cut process must be wetted in accordance with standard glove bag procedures. When not wetting ACM during removal is allowed, the ACM must still be adequately wet during storage, transport, and disposal.

(11) *Containerization of Waste at Elevations Exceeding 10 Feet.* Excluding removal of acoustical, sprayed-on, troweled-on, or fireproofing ceiling materials, ACM must be containerized at the height of removal if the elevation of the material is 10 feet or greater from the ground or floor. Roofing

waste may be lowered to ground level in a closed cluster in a pan scale or similar equipment prior to packaging.

(12) *Containerization of Asbestos Waste.* Prior to removal from the regulated area, asbestos waste must be containerized in fiber-tight leak-proof packaging and properly labeled, in accordance with OSHA 29 CFR, Part 1926.1101 (effective August 10, 1994).

(a) Friable asbestos waste and asbestos-containing cementitious materials removed from inside of occupied facilities. Friable asbestos waste that does not contain components with sharp edges must be adequately wetted and then containerized in two polyethylene bags with a 6-mil minimum thickness for each bag. Bags shall be then individually sealed in a fiber-tight manner by first removing air from the bag, and then twirling the open end of the bag sufficiently enough to permit the twirled end to be folded over and wrapped securely with duct tape. Fiber-tight drums may be used in lieu of bags to package waste. Liners inside dumpsters and roll-offs cannot be used to meet this containerization requirement.

If, however, the configuration or shape of the asbestos waste is such that the containerization of the asbestos waste in bags is not feasible, then it shall be adequately wetted and thoroughly wrapped in a minimum of two layers of 6-mil polyethylene sheeting with all joints, seams, and overlaps sealed in a fiber-tight manner. Asbestos-containing cementitious materials removed from inside of occupied facilities must be containerized in accordance with this section. All friable asbestos shall be disposed of in a landfill licensed to accept friable asbestos waste.

(b) Nonfriable asbestos waste may be packaged as set forth in Section 7(A)(12)(a) of this rule. At a minimum, nonfriable waste shall be wetted and containerized in leak-proof containers for delivery to a landfill that is licensed to accept nonfriable asbestos waste. Shredding, crushing, or any other form of volume reduction prior to placement in the landfill will render nonfriable asbestos waste subject to the containerization requirements set forth in Section 7(A)(12)(a) of this rule.

(c) Once visually clean, work area polyethylene sheeting must be removed and disposed of, either as asbestos or nonasbestos waste.

(13) *Onsite Storage of Asbestos Waste.* This is defined to be storage at the street address of the abatement site; it is subject to the following requirements:

(a) Waste must be stored in a secure container or area accessible to authorized persons only.

(b) Waste packaging must be free of visible debris prior to placement in the storage area or storage container.

(c) Waste must not remain onsite longer than 5 days following completion of asbestos abatement activities.

(d) Waste must be tracked by written documents, such as bill of lading or manifest, evidencing the current location of the waste at any time prior to final disposal.

Personal Decontamination Requirements

An individual must decontaminate prior to exiting the regulated area. Personal decontamination shall be achieved by removing all clothing and footwear except a bathing suit, if worn, and thoroughly showering with soap and water at a contiguous or remote decontamination facility. Respirators must be worn into the shower unit, and washed and cleaned as part of the decontamination procedure.

Clothing or footwear used or worn in the regulated area must not leave the equipment room unless containerized for reuse inside a regulated area, cleaned, or disposed of. Clothing worn under protective suits and footwear must be designated for asbestos use only, be easily identifiable by sight, and be permanently marked or labeled (minimum 3/4-inch lettering) as "Asbestos Clothing" on the outside of the clothing or footwear. Containerized clothing or footwear must be opened only inside a regulated area, excluding the clean room.

Only impervious materials such as rubber, polyethylene, etc. may be cleaned in the decontamination facility. Clothing and other pervious materials, such as leather boots, must be cleaned according to OSHA requirements.

Personal decontamination requirements for use with a remote decontamination unit are as follows:

(i) Each worker and/or supervisor shall be provided with appropriate personal protective and respiratory equipment as required by OSHA 1926.1101 (effective

August 10, 1994), including but not limited to a half-face respirator (at a minimum) and a protective suit (e.g., tyvek) or designated asbestos clothing. Wearing a bathing suit underneath disposable full-body suits is allowed.

(ii) Each worker and/or supervisor shall don his or her appropriate personal protective equipment in a manner such that the protective suit (e.g., tyvek) can be removed while the respirator is still being worn.

(iii) Before proceeding to and entering into a regulated area, each worker and/or supervisor shall remove all street clothes and footwear in the clean room of the remote decontamination facility.

(iv) Workers and/or supervisors wearing "Asbestos Clothing" as provided in Section 7(A)(14)(b) of this rule shall change into their asbestos clothing in the equipment room, first donning their respirator in the clean room, before proceeding to the equipment room. In the equipment room they shall place a clean protective suit and booties over their asbestos clothing and shoes before proceeding to the regulated area. The protective suit and booties can then be removed once inside the regulated area.

(v) Each worker and/or supervisor shall be provided with a clean, unused suit to carry to the regulated area. This suit shall be left at the designated egress point into the regulated area to be used when exiting from the regulated area.

(vi) Exiting the regulated area. Each worker and/or supervisor shall first remove all visible debris from the protective suit and then shall put the new clean unused suit that was left at the designated egress point to the regulated area entryway over his or her contaminated suit before proceeding to either another work area or the decontamination facility equipment room where both suits shall be removed in unison.

Respirators shall not be removed during this procedure and shall be worn at all times until they are removed in the shower room, during showering, as part of the decontamination process that must be performed at the conclusion of the day's shift or at any scheduled break period during the shift, including but not limited to the lunch break.

Equipment and Waste Decontamination Requirements

All equipment, supplies, and materials, including properly containerized waste material, work area ventilation units, HEPA vacuums, vacuum hoses, water hoses, extension cords, ladders, and so on, must be completely decontaminated and free of visible debris before removal from containment.

Where the size and/or shape of the equipment, supplies, and materials is such that decontamination is neither possible nor feasible (e.g., wood), then the object shall be properly containerized or wrapped in a minimum of two layers of fiber-tight 6-mil polyethylene sheeting for disposal or reuse in an active containment, and cleaned of visible debris prior to removal from the regulated area.

Decontamination of Work Area Ventilation Units

A work area ventilation unit must have the exterior filter(s) removed, immediately wetted, and disposed of as asbestos waste before the unit is taken out of containment. When the internal filters, including HEPA, of a work area ventilation unit are not changed upon project completion, the unit intake(s) and exhaust(s) must be wrapped fiber-tight with a minimum of two layers of 6-mil polyethylene sheeting before removal from containment. Internal filters must be removed inside an active containment, adequately wetted immediately upon removal from the unit, and disposed of as friable asbestos waste.

HEPA vacuums must be emptied in an active containment that includes an operating work area ventilation system and decontamination facility. Sweeping of dry ACM is prohibited, and all removed ACM must be containerized by the end of each workday.

Existing Asbestos-Containing Debris in the Regulated Area

Visible asbestos-containing debris that is present on surfaces on which the contractor will place polyethylene sheeting to establish the regulated area shall be cleaned up prior to conducting setup, removal, or repair activities. The Asbestos Design Consultant must demarcate the regulated area, incorporate into the design any existing debris within the regulated area, and consider debris part of the abatement activity. The Asbestos Abatement Contractor must remove existing asbestos-containing debris on all surfaces and components within the regulated area as part of abatement activity.

All visible dust and debris must be removed from the regulated area. The regulated area must be cleaned and dry, and surface coatings must not be applied to any surface within the regulated area, prior to conducting the first visual evaluation and subsequent air clearance sampling.

Teardown

Following the initial visual evaluation and receipt of acceptable air clearance sampling results, the contractor shall remove the containment, critical barriers, and decontamination unit from the work site. The contractor shall clean up any visible dust or debris resulting from teardown activities prior to the final inspection after removal of containment in accordance with Section 8(B)(3) of this rule.

An asbestos abatement activity is not considered complete and acceptable for regulated area release until initial visual evaluation standard(s), standard(s) for visual evaluation at the final inspection, and final air clearance standards(s) (if applicable) have been met as set forth in these rules. All applicable work practice requirements set forth in this rule must continue to be implemented until the project is complete.

Alternative Work Practice Requirements for Demolition Activities

Except as allowed in Section 7(B)(3) of this rule, demolition of a building that contains ACM shall not commence until all ACM has been removed. A certified Design Consultant must specify materials to be removed and/or left in place for demolition activities conducted pursuant to this rule.

Removal of ACM from a building being demolished must comply with the work practice requirements set forth in this section except that static clearances are allowed when dirty or dusty conditions not related to asbestos abatement activities exist outside the regulated area that likely will result in count overloads to filter media. Static clearances are required when aggressive methods are not feasible.

For buildings where demolition is scheduled to occur within 6 months of the asbestos abatement project and general access to the building will be restricted, the regulated area does not need to include work area containment as set forth in Section 7(A)(2)(a) of this rule. Air ventilation unit(s) shall be placed within the regulated area adjacent to active removal activities. If this is not feasible, the asbestos abatement contractor shall submit a nonstandard variance to the Department stating the reason(s) for not placing unit(s) within the regulated area.

Prior to beginning a partial demolition project where asbestos will be impacted, the remaining section(s) of the building adjacent to the asbestos demolition must be isolated with a minimum of two layers of fiber-tight 6-mil polyethylene sheeting, and HVAC equipment in or passing through the demolition area must be isolated with the 6-mil polyethylene sheeting.

Demolition by Large Equipment of Buildings Containing Intact Flooring Materials

Intact asbestos-containing flooring does not require removal prior to demolition by large equipment (e.g., bulldozers with rakes, top loaders, backhoes, skid loaders/bobcats, hydraulic excavators, cranes with wrecking balls, clamshells, or buckets, and other similar machinery), provided that the following alternative work practices, as cited in (a) through (g) are implemented. Contractors performing these operations must be licensed by the Department.

Employees performing the work must be certified by the Department. OSHA training shall be documented and copies of these training certificates shall be at the asbestos work site for any supervisors and workers certified under the certification-by-rule provision of Section 5(A)(2) of this rule. A licensed consultant must document in writing that all flooring material is intact and that the contractor performing the demolition activity is licensed in accordance with the requirements of this section.

(a) A regulated area must be established.
(b) The project must be conducted in a manner that minimizes the release of asbestos fibers. All necessary and appropriate measures must be taken to ensure that release of asbestos fibers is minimized. The ACM must be kept wet at all times during the demolition, onsite

storage, transportation, and disposal activities. If visible emissions are observed during demolition of an area with asbestos-containing materials, work shall cease until engineering controls are in place to prevent such visible emissions.

(c) Employees within the regulated area must be trained consistent with OSHA 29 CFR, Part 1926.1101 (effective August 10, 1994). Training shall be documented, and the training documentation shall be made available immediately to the Department at the work site.

(d) Employees within the work area must wear appropriate personal protective equipment, including a minimum of a half-face respirator equipped with HEPA filters and full-body coverings, including hand and foot coverings.

(e) Asbestos waste must be containerized in leak-proof transport vehicles and securely covered during transport. Waste may be segregated into asbestos waste and nonasbestos waste as needed to meet disposal facility requirements. Unless the asbestos-containing flooring is separated from all other demolition debris generated by the demolition activity, all of the demolition debris from the activity must be disposed of as asbestos waste.

(f) Disposal of asbestos-containing demolition debris must occur at a landfill licensed to accept construction/demolition debris or asbestos waste. The nonfriable asbestos waste must be containerized in accordance with Section 7(A)(12)(b) of this rule.

(g) A visual evaluation of the regulated area must be performed in accordance with Section 8(B) of this rule prior to the release of the regulated area.

Additional Requirements for Enclosure and Encapsulation Activities

Enclosure and encapsulation are considered to be asbestos abatement activities for purposes of these rules and are subject to the work practice requirements of Section 7(A) of this rule and the following:

(1) *Enclosure.* Enclosures must be labeled or identified in the permanent building records to indicate the presence of ACM within them.

(2) *Encapsulation.* Liquid-penetrating encapsulants must be applied with airless spray equipment, brushes, or rollers and in accordance with the manufacturer's recommendations. Liquid encapsulants must not be applied to damaged or deteriorated ACM except to seal pipe ends during a glove bag operation or during repair operations. A bridging encapsulant (including rewettable cloth and a pliable heat-resistant mastic) only shall be applied over damaged thermal system insulation.

Roofing and Exterior Asbestos-Containing Cementitious Products Projects

This section establishes work practice requirements for roofing projects that involve more than 105 square feet of asbestos-containing roofing materials that are removed by mechanical roof saws or cutters. It also establishes work practice requirements for exterior asbestos-containing cementitious products removal projects. See Figure 9.1 for an example of the removal of asbestos roofing panels.

Contractors performing these operations must be licensed by the Department. Employees performing the work must be certified by the Department. OSHA training shall be documented and copies of these training certificates shall be at the asbestos work site for any supervisors and workers certified under the certification-by-rule provision of Section 5(A)(2) of this rule.

Figure 9.1 Removal of asbestos roofing panels.

This documentation must be made immediately available to the Department upon request.

Roofing projects involving the removal of asbestos-containing roofing materials by mechanical roof saws or cutters are subject to the following requirements:

(a) At least one onsite employee must be trained as a competent person.

(b) A regulated area must be established, except that containment, a ventilation system, critical barriers, and a waste load-out area are not required, and a remote decontamination unit is allowed.

(c) HVAC intakes and exhausts inside the work area must be isolated and sealed with fiber-tight 6-mil polyethylene sheeting.

(d) Workers performing the cutting and cleaning operations must wear appropriate personal protective equipment as prescribed by 29 CFR 1926.1101 (effective August 10, 1994), including, at a minimum, half-faced respirator equipped with HEPA filters and full-body coverings. Figure 9.2 shows an example of personal protective equipment for roofing projects.

(e) Workers must comply with the personal decontamination requirements of Section 7.A(14) of this rule.

Figure 9.2 Asbestos worker in protective clothing while removing roofing.

(f) The roof cutter must be equipped with an operational blade cover.

(g) The roof cutter must be continuously misted during operation.

(h) The tailings from the roof cutter shall be kept wet and cleaned up by HEPA vacuuming or wet-wipe techniques.

(i) The roof area being abated or cut shall be kept wet at all times.

(j) The tailings, and any other friable asbestos waste, shall be containerized and stored in accordance with Sections 7.A(11) and 7.A(12)(a) of this rule by the end of each work day.

(k) The nonfriable waste, including the small sections of cut (not torn) built-up roofing, shall be containerized in accordance with Section 7.A(12)(b).

(l) Prior to deregulating the regulated area, the regulated area shall be cleaned and meet the visual evaluation standard of Section 8.B(1) of this rule.

(m) Visual evaluation of each regulated area is required, but air clearances are not required.

Visual evaluations on roofs shall be conducted by an Asbestos Project Supervisor, Air Monitor, or OSHA-trained competent person, and are not subject to the "Conflict of Interest" provisions of Section 2.G of this rule.

Exterior asbestos-containing cementitious products. Work with siding or other projects involving exterior asbestos-containing cementitious products ("transite") are subject to the following requirements:

(a) At least one onsite employee shall be trained as a competent person.

(b) A regulated area shall be established, except that a containment, a ventilation system, critical barriers, and a waste load-out area are not required, and a remote decontamination unit is allowed.

(c) Appropriate personal protective equipment shall be utilized while inside the regulated area, consisting of half-face respirator equipped with HEPA filters and full-body coverings, including head and foot coverings.

(d) Workers shall comply with the personal decontamination requirements of Section 7(A)(14) of this rule.

(e) The material shall be thoroughly wetted before and during the removal to ensure prompt wetness (especially back side) and ensure that it stays wet during removal, storage, and transport to the landfill.

(f) The material shall be removed as whole as possible and carefully lowered, not dropped, to the ground.

(g) The cementitious product shall be containerized in accordance with Section 7(A)(12)(b).

(h) Prior to deregulating the regulated area, the regulated area shall be cleaned and meet the visual evaluation standard of Section 8.B(1) of this rule.

(i) Visual evaluation of each regulated area is required, but air clearances are not required. Visual evaluations on cementitious siding shall be conducted by an Asbestos Project Supervisor, Air Monitor, or OSHA-trained competent person, and are not subject to the "Conflict of Interest" provisions of Section 2.G of this rule.

Glove Bag Operations and Wrap-and-Cut Projects

These projects involve the removal of components covered with thermal system insulation that utilizes "wrap-and-cut" methods; removal or repair of asbestos-containing materials that involve use of multiple noncontiguous glove bags that are no larger than 60 inches by 60 inches; removal or repair, using contiguous glove bags, that involve a total of no more than 30 linear feet of ACM on a single pipeline; or any amount of asbestos-containing materials that can be removed within 10 glove bags for pipelines running parallel to each other.

Glove bag and wrap-and-cut projects are subject to Section 2(G) of this rule, Conflict of Interest, and Section 8(B) of this rule, Release of the Regulated Area requirements as determined by the total amount of ACM to be removed during the project as indicted on the notification form in Section 12 of this rule. A certified Asbestos Inspector or Design Consultant shall evaluate the component(s) to determine that the component(s) is intact and not likely to release fibers during removal. The determination shall be in writing and shall be recorded in the project design.

(1) *Glove Bag Operations*

(a) Remove or cover with a single layer of 6-mil poly movable objects within the proposed regulated area.

(b) Cover immovable objects within the proposed regulated area with a single layer of 6-mil poly.

(c) Establish critical barriers (where applicable) and demarcate the regulated area with barrier tape marked "Asbestos Hazard."

(d) Establish a decontamination facility contiguous with the regulated area.

(e) Conduct glove bag removals using recognized glove bag removal techniques.

(f) Conduct a visual evaluation by a certified Air Monitor of the completed glove bag removal before the glove bag is removed per Section 8(B)(1)(a),(c),(g), and (j) of this rule.

(g) Remove glove bags from pipes/components.

(h) Remove the glove bags from the regulated area.

(2) *Wrap-and-Cut Operations Requiring the Use of Glove Bags*

(a) Perform work area preparation as described in Section 7(E)(1)(a–d) of this rule.

(b) Wet the pipes/component with water.

(c) Wrap the pipes/component with two layers of 6-mil polyethylene sheeting, overlapping the seams and securing with duct tape creating a fiber-tight container.

(d) Conduct glove bag removals at appropriate intervals using recognized glove bag removal techniques.

(e) Conduct a visual evaluation by a certified Air Monitor of the completed glove bag removal and wrapping operation before the glove bag is removed per Section 8(B)(1)(a), (c), (g), and (j) of this rule.

(f) Remove glove bags and cut exposed pipe using appropriate method.

(g) Remove the glove bags and wrapped pipes/components from the regulated area.

(3) *Work Practice Requirements for Wrap-and-Cut Projects Not Requiring the Use of Glove Bags*

(a) Perform work area preparation and component removal wrapping as described in Section 7(E)(1)(a–d) of this rule.

(b) Conduct a visual evaluation by a certified Air Monitor of the completed wrapping operation before the pipe/component is cut.

(c) Cut the component(s) as applicable.

(d) Remove the wrapped pipes/components from the regulated area.

Nonstandard Work Practice Requirements

Nonstandard work practices may be permitted when the standard procedure is not practicable, not feasible, not safe, or when a cost-saving alternative exists and the proposed nonstandard work practice adequately protects both human health and safety, as well as the environment, from exposure to asbestos hazards. Nonstandard work practices must be developed by a certified Design Consultant and must be sent in writing to the Department with the original notification form unless unforeseeable conditions occur during a project that warrant a request at that time.

The nonstandard work practice must present clear and convincing evidence that the asbestos project is distinctive in some way and the proposed alternative(s) to required work practices will comply with the intent of state law and these rules. Where applicable, the Design Consultant submitting the nonstandard work practice must notify the Asbestos Design Consultant who prepared the original project design for the project of the nonstandard work practice(s) submission to the Department.

This notification must be concurrent with the nonstandard work practice submission to the Department. Nonstandard work practices require written authorization from the Department prior to implementation. The Department will respond to nonstandard work practice requests within 5 working days of receipt and will indicate whether the proposal is authorized or not, and if not, why not.

The Department can revoke a nonstandard work practice approval whenever additional information is obtained or a change in project conditions occurs. When given a nonstandard work practice authorization, the abatement contractor still must comply with all other applicable provisions of this rule and other state and federal rules and regulations.

Waste Shipment Records

For all asbestos-containing waste material transported off the facility site, the operator shall maintain waste shipment records and include the following information:

(a) The name, address, and telephone number of the waste generator

(b) The name and address of the local, state, or EPA regional office responsible for administering the asbestos NESHAP program

(c) The approximate quantity of waste, measured in cubic meters (cubic yards)

(d) The name and telephone number of the disposal site operator

(e) The name and physical site location of the disposal site

(f) The date transported

(g) The name, address, and telephone number of the transporter(s)

(h) A certification that the contents of this consignment are fully and accurately described by proper shipping name and are classified, packed, marked, and labeled, and are in all respects in proper condition for transport by highway according to applicable international and government regulations

Provide a copy of the waste shipment record, as described earlier, to the disposal site owners or operators at the same time as the asbestos-containing waste material is delivered to the disposal site. For nonfriable asbestos waste only, a copy of the Maine Department of Environmental Protection Non-Hazardous Waste Transporter Manifest for Category A waste may be used to meet this waste shipment records requirement. The licensed nonhazardous waste transporter is required to complete and maintain this manifest form and to provide a copy to the disposal facility; asbestos abatement contractors may request a copy of this form from the transporter.

In instances in which a waste shipment record, signed by the owner or operator of the designated disposal site, is not received by the waste generator within 35 days of the date the waste was accepted by the initial transporter, the operator must contact the transporter and/or the owner or operator of the designated disposal site to determine the status of the waste shipment.

Report in writing to the local, state, or EPA regional office responsible for administering the asbestos NESHAP program for the waste generator if a copy of the waste shipment record, signed by the owner or operator of the designated waste disposal site, is not received by the waste generator within 45 days of the date that the waste was accepted by the initial transporter. The report must include a copy of the waste shipment record for which a confirmation of delivery was not received and a cover letter signed by the waste generator explaining the efforts taken to locate the asbestos waste shipment and the results of those efforts.

There must be a copy of all waste shipment records, including a copy of the waste shipment record signed by the owner or operator of the designated waste disposal site, retained for at least 2 years. Information must be furnished upon request, and made available for inspection by the Department, all records required under this section.

▶ MONITORING AND RELEASE REQUIREMENTS FOR A REGULATED AREA

Asbestos abatement activities are subject to the following air and project monitoring requirements:

(1) *Air monitoring,* in the form of air clearance sampling, must be conducted for all asbestos abatement activities. A project-specific air monitoring plan, including the mandatory air clearance sampling and optional additional area monitoring, must be developed and used for each project and the plan must be part of the project design. Any change(s) to the air monitoring plan must be made by a certified Asbestos Air Monitor or Asbestos Abatement Design Consultant and must become part of the project documentation. The air monitoring plan must be designed in accordance with these rules by an Air Monitor or Asbestos Abatement Design Consultant who has an independent business relationship with the asbestos abatement company performing the project whenever independent air clearances are required.

(2) *Project monitoring* is not required by these rules but may be required due to contractual or other arrangement or specification. If an asbestos abatement activity is monitored, then written records of all project monitoring activities shall be maintained at the project site for the duration of the project and shall become part of each owner's and operator's official records for that project. This project monitoring documentation must be provided to the building owner within 6 months of project completion.

(3) *Air samples* collected for air clearance sampling must be analyzed by a Department-licensed Asbestos Analytical Laboratory as follows:

    (a) Each air sample collected shall be analyzed, as per the licensing section of this rule, by phase contrast microscopy (PCM), transmission electron microscopy (TEM), or another Department-approved method.

    (b) For air sample analyses, results obtained from the TEM methodology, described in Appendix A to Subpart E, 40 CFR, Part 763, shall be considered definitive when results differ.

Release of Regulated Area

A regulated area may not be released from the contractor's control until no visible debris remains in the regulated area, visual evaluations are completed, air clearance sampling in accordance with this section is conducted, and clearance standards are met. Visual evaluations and air clearances for an asbestos abatement project involving more than 100 linear/square feet, or any combination thereof, of ACM must be performed by an Asbestos Consultant. Visual evaluations and air clearance sampling must be conducted by a certified Asbestos Air Monitor employed by an Asbestos Contractor, in-house Asbestos Abatement Unit, or Asbestos Consultant.

The following procedures shall be performed sequentially and documented:

(1) After final abatement activities, including final cleaning and removal of equipment, supplies, and waste, and prior to removal of any layer of containment (if applicable) or glove bag and before conducting air clearance sampling, a visual evaluation of the regulated area shall be conducted, as specified next, to ensure that there is no visible dust or debris in the regulated area, including the containment. The regulated area must be completely dry such that there is no visible evidence of water in the regulated area prior to conducting the visual evaluation. The individual conducting the visual evaluation where there is a containment or glove bag shall

    (a) Inspect the decontamination facility (including remote decontamination facilities) and waste load-out unit (where applicable) to ensure there is no visible dust or debris present.

(b) Enter the regulated area/containment where the asbestos abatement activity was performed.

(c) Get close enough to see and touch the surfaces from which the ACM was removed or on which other abatement operations were performed.

(d) Inspect the surfaces from which the ACM was removed.

(e) Observe and touch the substrate.

(f) Examine the permanent features of the regulated area such as walls, conduits, pipes, ceiling tile grid bars, ducts, etc.

(g) Examine the floor, walls, and other surfaces of the regulated area.

(h) Examine decontamination and waste load-out facilities.

(i) Examine places where the polyethylene sheeting may have fallen away from the walls or partitions.

(j) Examine the polyethylene floor coverings to determine whether visible debris and/or contaminated water may have seeped through the plastic layers. If debris is observed, the regulated area must be cleaned and another visual evaluation conducted. Project documents must reflect the recleaning activity(ies). The Air Monitor must ensure that the regulated area is free of visible debris before conducting air clearance sampling. The Air Monitor must document in the daily project log the time that the regulated area was determined to be free of visible debris so that air clearance sampling could commence.

(2) Air clearance sampling must be performed and documented in accordance with this rule. Air clearance sampling is subject to the following requirements:

(a) The containment must be completely dry such that there is no visible evidence of water in the regulated area prior to conducting air clearance sampling.

(b) Immediately prior to conducting air clearance sampling, the Air Monitor shall implement aggressive sampling by sweeping the walls, ceiling, and floor with the exhaust of a minimum 1-horsepower leaf blower. Stationary fans shall then be placed in locations that will not interfere with air sampling equipment. Fan air shall be directed to the ceiling. One fan shall be used for each 10,000 square feet of regulated floor area and shall run throughout the air clearance sampling event. Aggressive sampling is not required when a regulated area is in a dirt crawl space or when dirty or dusty conditions not related to asbestos abatement activities exist outside the regulated area and will result in rendering filter media unreadable. Static air clearance samples are required when aggressive sampling is not appropriate.

(c) The minimum number of air clearance samples is as follows:

    (i) 2 samples for activities that contain less than 100 linear/square feet or any combination thereof of ACM

    (ii) 3 samples for activities that contain more than 100 but less than 1,000 linear/square feet or any combination thereof of ACM

    (iii) 5 samples for activities greater than 1,000 linear/square feet or any combination thereof of ACM

(d) PCM air clearance samples must contain at least 2,452 liters of air and the sampling flow rate must not exceed 16 liters of air per minute.

(e) Notwithstanding the preceding, for asbestos abatement projects conducted in schools, the number and flow rate of the air clearance samples must be in accordance with the requirements of 40 CFR, Part 763, Subpart E.

(f) Air clearance samples shall be analyzed as follows:

    (i) For asbestos abatement projects in schools that impact more than 160 square feet or 260 linear feet of asbestos-containing material, air clearance sample analysis must be in accordance with the requirements of 40 CFR, Part 763, Subpart E, Appendix A (effective December 14, 1987).

    (ii) For all other projects, air clearance sample analysis must be in accordance with Appendix A of 40 CFR (referenced before), the most current version of the National Institute for Occupational Safety and Health (NIOSH) Methods 7400 or 7402, as applicable the OSHA Reference Method Asbestos Standard for General Industry, 29 CFR 1910.1001 Appendix A (effective date July 20, 1986) for personal air samples, or another Department-approved analytical method.

    (iii) The total fiber count of each of the air clearance samples collected inside the regulated area must be less than or equal to 0.010 f/cc (fibers per cubic centimeter) of air (as analyzed by phase contrast microscopy), or must be less than or equal to $70/\text{mm}^2$ (structures per square millimeter) by transmission electron microscopy (TEM), or must be below the clearance criteria for another Department-approved method(s) to release the work area.

(g) Failures of air clearance sampling (not meeting the clearance criteria of 0.010 f/cc or 70 s/mm²) require that the asbestos abatement contractor:

   (i) Wet wipe and HEPA vacuum the entire regulated area. Potentially contaminated make-up air may be prefiltered and/or excluded from entering the regulated area prior to recleaning the entire regulated area.

  (ii) Resample and reanalyze either by NIOSH 7400 or 7402 (TEM) or AHERA TEM 40 CFR, Part 763 (effective October 30, 1987), and school projects must follow the AHERA sampling and analysis protocols.

  (iii) Repeat wet wiping, HEPA vacuuming, and resampling until the air clearance standards of this section are met.

(3) Immediately upon completion of removal of the containment of the regulated area, an Asbestos Air Monitor or Asbestos Project Supervisor must visually inspect all surfaces within the regulated area for visible debris. If there is no containment, then the visual evaluation of the regulated area shall be consistent with Section 8.B(1) of this rule. If visible debris is observed, the regulated area must be cleaned by HEPA vacuum or wet methods until there is no visible dust or debris present. This final inspection must be documented in the daily project log. This documentation must include a statement that the regulated area was clear of visible debris and the name and signature of the person conducting this final inspection.

▶ PERMIT-BY-RULE FOR ASBESTOS WASTE STORAGE FACILITIES

Prior to disposal off the site of generation, asbestos waste may be stored by a business or public entity in quantities greater than 1 cubic yard only at a Department-licensed Asbestos Waste Storage Facility (AWSF) permitted under the provisions of this section. An AWSF must not be located within 500 feet of any public or private school, daycare, or preschool, or any other such building utilized for the education of students in grades K–12.

A variance of this requirement may be granted by the Department under extenuating circumstances where preexisting storage facilities are currently located within 500 feet of educational facilities described in this section and no other feasible storage alternative exists. Also, the ASWF must be located on property in which the licensee has title, right, or interest.

A business or public entity that operates or intends to operate an AWSF must apply to the Department for a permit-by-rule to operate an AWSF at least 60 days prior to the operation of the AWSF. A licensee must notify the Department at least 14 days in advance whenever it will cease operating its permitted AWSF. Moving an AWSF to a new location requires submitting a new permit-by-rule application except that there is no 14-day waiting period prior to using the AWSF after the Department has approved the application.

A business or public entity that obtains an AWSF permit-by-rule pursuant to this section need not obtain a solid waste storage facility permit pursuant to Maine's *Transfer Stations and Storage Sites for Solid Waste Rules*, 06-096 CMR 402 of the Maine Solid Waste Management Regulations.

For approval as an AWSF, the following information must be submitted by the applicant on a Department form:

(1) Applicant's name, address, telephone number, contact person, and responsible person with signature
(2) Anticipated annual asbestos waste volume based on previous operational data or any other relevant data
(3) General description of the AWSF including its location
(4) Site plans, including but not limited to
    (a) Facility boundaries
    (b) Location of the AWSF
    (c) Site security systems
    (d) Fences and gates
    (e) Existing structures within 500 feet of the AWSF boundary lines with owner(s) name(s) and address(es)
(5) Most recent full-sized U.S. Geological Survey topographic map (7-½ minute if available) or other similar map detailing the property and an AWSF Operations Manual.

AWSF Operations Requirements

The AWSF operator must prepare and maintain an operations manual regarding the day-to-day operations of the AWSF. The AWSF operator is responsible for ensuring that persons involved in the day-to-day operations of the AWSF are familiar and comply with the operations manual and the requirements of this section.

The AWSF operations manual must be submitted to the Department with the permit-by-rule application and must include procedures to ensure the following requirements are met:

(a) The AWSF must be locked and impact resistant.

(b) The waste must be packaged in a minimum of two layers of 6-mil polyethylene sheeting and be fiber-tight, as required by these rules.

(c) Asbestos waste must be adequately wet during storage.

(d) AWSF labeling (placarding) and asbestos waste labeling must be in accordance with OSHA's 29 CFR 1926.1101 (effective August 10, 1994).

(e) Containerized asbestos waste must not be dropped or thrown further than 3 feet, and caution and due care must be utilized during asbestos waste handling.

(f) Asbestos waste that is leaking or improperly packaged must be immediately repackaged. Dry asbestos waste must be immediately repackaged under controlled abatement conditions complete with negative pressure ventilation.

(g) Site security procedures must be established to prevent unauthorized persons from entering the AWSF.

(h) Asbestos waste must be removed from the site and transported to a licensed disposal facility at least once per year.

▶ APPROVAL OF TRAINING COURSES

A training course used by an individual to fulfill certification requirements must be approved by the Department or by another governmental agency that has a reciprocal arrangement with the Department. Individuals are responsible for ensuring that a training course is approved before taking the course. Courses approved by another state or the U.S. EPA may or may not be approved by the Department depending on the length, content, and presentation of the course and the qualifications of the Training Provider. Department approval of a training course expires after 1 year.

An application for training course approval must be submitted to the Department at least 30 days prior to the first scheduled training course date on forms approved by the Department. Training course curriculum and operational procedures used by the Training Provider must adhere to the U.S. EPA Asbestos Model Accreditation Plan; Interim Final Rule, 59 FR 5236-5260 (effective April 4, 1994); and Appendix C to Subpart E of 40 CFR, Part 763, "Asbestos-Containing Materials in Schools" rule (effective December 14, 1987). Initial and refresher training courses are subject to the Interim Final Rule.

If a training course is not referenced in Appendix C to Subpart E of 40 CFR, Part 763, "Asbestos-Containing Materials in Schools" rule, the Department shall determine adequacy of the course by reference to industry standards, training offered by other trainers, the requirements of this section, and other material requested from the applicant.

Course Content

Course content for the 16-hour Asbestos Air Monitor course that may be used in conjunction with a supervisor course to attain certification as an Asbestos Air Monitor must include at a minimum:

(a) Air sampling procedures, strategies, and equipment
(b) Pump calibration techniques
(c) Air sampling calculations
(d) Limitations of air sampling
(e) Appendix A of "Asbestos-Containing Materials in Schools" rule, 40 CFR, Part 763 (effective October 30, 1987), including NIOSH methods
(f) Response action clearances
(g) Required recordkeeping and documentation
(h) Applicable state and federal rules and regulations, including but not limited to this rule, NESHAP, EPA, and OSHA
(i) Quality assurance procedures
(j) Duties and responsibilities of a certified Asbestos Air Monitor, as set forth in this rule and other applicable rules and regulations
(k) Project monitoring techniques
(l) Visual evaluation techniques
(m) Visual clearances as per this rule and the American Society for Testing and Materials (ASTM), Method E 1368-90
(n) Reading and interpreting specifications and drawings

Course content for an initial Asbestos Air Analyst must include techniques and procedures for quantification of fibers in air samples and at a minimum cover all topics presented in the former NIOSH Course #582, "Sampling and Evaluation of Airborne Asbestos" or equivalent.

The Asbestos Air Monitor refresher course shall be 4 hours. An 8-hour OSHA Worker and 12-hour OSHA Competent Person course content must include subject matter required by OSHA, pursuant to 29 CFR 1926.1101 (effective August 10, 1994), as well as proper clean-up techniques, visual evaluation of the work area, proper waste packaging, and work stoppage criteria.

Application Requirements for Approval of Training Courses

An applicant for approval of a training course must provide the following:

(1) The name, address, telephone number, and license number of the DEP-licensed Training Provider conducting the training course, including, if applicable, any other name under which the Training Provider is known

(2) The name of the training course for which approval is sought

(3) A course curriculum detailing specific topics to be covered along with allotted topic times

(4) A copy of the training course manual along with all printed material

(5) A description of the teaching methods to be utilized, including but not limited to a description of audiovisual aids

(6) A description of the "hands-on" facility to be utilized, including but not limited to protocol for instruction and ensuring direct contact with actual situations encountered in the field of study

(7) A statement that the student-to-instructor ratio for the hands-on portions will be no greater than 10 to 1

(8) A description of the equipment that will be utilized in classroom lectures and in hands-on training

(9) The names, background, qualifications, and training and experience of the primary instructor and the names and qualifications of all secondary instructors providing the training

(10) A copy of the uniquely numbered certificate of course completion, which must contain the following:
  (a) The name of the student
  (b) The name of the training course completed
  (c) The date(s) and location of the training course
  (d) The certificate's date of expiration
  (e) A statement that the student passed the examination
  (f) The name, address, and telephone number of the Training Provider
  (g) The length of the course in hours
  (h) The name and signature block of the primary instructor for the course
(11) Other information necessary to determine the adequacy of the training course content and presentation
(12) Any other state or jurisdiction under which the course may be approved

Renewal Application Requirements for Approval of Training Courses

The applicant must demonstrate it meets the following requirements for renewal of approval of training courses:

(1) Submission of a renewal application
(2) Submission of any changes in the course content, curriculum, instructors, and exam questions and procedures
(3) Compliance with the following:
  (a) Standards of Conduct set forth in Section 5(F) of this rule
  (b) Maintaining the training/facility standards and course content set forth in the original license application
(4) Written response to Department training course audit findings addressing how each finding has been incorporated into course curricula/materials or facility standards noted in the audit

Successful Course Completion

Successful completion of a training course requires that a student attend at least 90% of course time and achieve a score of 70% or greater on the course exam. Except refreshers, primary instructors cannot be considered students for consideration of successful course completion.

Examinations

Exams are required for all training courses conducted pursuant to this rule. Passing an exam is achieved when a student achieves a score of 70% or higher. Failure of an exam requires:

(a) That a student retake the initial or refresher training course exam once.

(b) That a student not passing the second exam must attend 8 hours of remedial training prior to taking the initial or refresher training course exam again.

(c) That a student failing the third exam after taking the remedial training must take the initial or refresher training course again along with the training course exam.

(d) The Training Provider will grade the exam and communicate the course results to the Department within 5 working days. The course results must include the type of course, the names, Social Security numbers, and exam scores of all participants.

Initial Exam

Initial training courses require a final exam, which shall be provided and administered by the Training Provider. The Department will provide the Training Provider with a matrix describing the topics on the exam, the number of questions required on each exam, and the approximate percentage of questions on each topic. The Department reserves the right to provide and/or administer final course exams. Alternate procedures, including but not limited to those arising out of reciprocal agreements with other states, must be approved in writing.

Refresher Course Exam

Exams developed and administered by a licensed Training Provider are required for annual refresher training courses. The exam shall be a minimum of 25 questions, shall be approved by the Department, and shall adequately address the refresher topics referenced in Appendix C, Interim Final Rule, 59 FR 5236-5260 (effective April 4, 1994) to Subpart E of 40 CFR, Part 763, "Asbestos-Containing Materials in Schools" rule (effective December 14, 1987), and other requirements listed in this rule.

Reciprocity

Reciprocity of successful completion of training courses is allowed by this rule. Individuals seeking certification in Maine must submit documentation to the Department affirming that the training they received was at least as stringent as the training course requirements of this rule. Any of the requirements of this section, as well as the requirements for licensure of Training Providers set forth in this rule, may be requested by the Department when determining adequacy of training courses.

Course Notification to Department

Training Providers must notify the Department in writing of asbestos training courses conducted pursuant to this rule within the geographic boundaries of the State of Maine, on forms approved by the Department and at least 10 calendar days prior to the start date of the course. Included in the notification shall be the location of the course offering, and if other than the business location of the Training Provider, sufficient information to demonstrate that the new location is adequate for the training and learning purposes of the course.

Courses not properly notified to the Department may not be approved by the Department. The Training Provider shall notify the Department via fax or phone of the cancellation of any course no later than 7:00 a.m. the day of the course.

Course Times

Training courses conducted pursuant to these rules shall be provided during normal business day hours, Monday to Friday, and no earlier than 7:00 a.m. and no later than 6:00 p.m. Training shall not occur on state holiday days. A course day must not be less than 6½ hours or exceed 8 hours in duration, excluding lunch and breaks. Other course days and times may be approved by the Department on a case-by-case basis. Courses must be completed within a 2-week period.

▶ REGULATIONS INCORPORATED BY REFERENCE

These rules attempt to be consistent with federal asbestos rules and regulations established by OSHA and the U.S. EPA. However, regulations established by OSHA and the EPA are inconsistent; therefore, these rules cannot be entirely consistent with both agencies. These rules are promulgated with the welfare of the general public and the environment as a highest concern and are intended to promote safe and proper asbestos management in the state of Maine. No provision of this rule shall be construed to preempt or supersede any other provision established under another statute or jurisdiction.

The following federal rules and regulations of the U.S. EPA are incorporated by reference herein as Maine Department of Environmental Protection, *Asbestos Management Regulations*, 06-096 CMR 425:

(1) "Asbestos-Containing Materials in Schools" rule; 40 CFR, Part 763, Subpart E (effective December 14, 1987)
(2) U.S. EPA Asbestos Model Accreditation Plan; Interim Final Rule, 59 FR 5236-5260 (effective April 4, 1994), Appendix C to Subpart E of 40 CFR, Part 763, "Asbestos-Containing Materials in Schools" rule (effective December 14, 1987)

Applicability
The applicability of each rule and regulation that is incorporated herein by reference is outlined in the following:

(1) Subpart E of the "Asbestos-Containing Materials in Schools" rule (40 CFR, Part 763). This rule is applicable to local education agencies (LEAs) in Maine. This federal regulation establishes ongoing inspection and management criteria for asbestos in schools and is not applicable to any other facilities in Maine.
(2) Interim Final Rule, 59 FR 5236-5260, Appendix C to Subpart E of the "Asbestos-Containing Materials in Schools" rule (40 CFR, Part 763). This rule establishes minimal training course and Training Provider requirements for persons seeking state certification and applies to all persons conducting asbestos abatement activities who may seek state certification in Maine, as well as Training Providers offering asbestos training courses.

Substitution of Definitions
The provisions of federal regulation that are incorporated by reference into this rule can be understood in terms of state law by making the following substitution in terms utilized therein:

"EPA" means the "Maine Department of Environmental Protection (DEPARTMENT)"; "Regional Administrator" and "Director" mean the "Board of Environmental

Protection or its designated representative"; and "Definitions" shall mean the definition of each term as it applies to the applicable law, regulation, or rule in which the definition is found. Defined terms are therefore specific to the regulation, law, or rule in which they are found and not meant to be generic in nature.

Conflict of Federal and State Law

Where the provisions and/or terms of the federal rules and/or regulations incorporated by reference into this rule differ from or are inconsistent with other terms and/or provisions of this rule, the term and/or provision of the more stringent rule shall apply.

▶ CONCLUSION

We are finally done with the example of EPA regulations for asbestos workers in the state of Maine. Always remember that when you are working with asbestos, you are subject to rules and regulations set forth by the EPA, OSHA, and local jurisdictions. The work is a complicated process when all of the restrictions are factored in. Many of the rulings overlap, but you cannot just assume that they do. It is your responsibility to be aware of all laws, rules, and regulations pertaining to the work that you will be doing.

Hazardous Materials Workers 10

What does becoming an asbestos worker entail? How difficult is it to get into the business? Does asbestos work pay well? Formal education beyond high school is not required, but government standards require specific types of on-the-job training. This training can be accomplished within weeks. Good job opportunities are expected in the future. The need to replace workers who leave the occupation creates ongoing demand. And, as regulations evolve and more and more work with asbestos is understood and done, the demand for certified workers should be steady for some time to come. There is a drawback. The working conditions can be hazardous.

▶ NATURE OF THE WORK

The workers who remove hazardous materials identify, remove, package, transport, and dispose of asbestos, radioactive and nuclear waste, arsenic, lead, and mercury—or any materials that typically possess at least one of four characteristics—ignitability, corrosivity, reactivity, or toxicity. These workers often respond to emergencies where harmful substances are present, and are sometimes called abatement, remediation, or decontamination specialists. Increased public awareness and federal and state regulations are resulting in the removal of hazardous materials from buildings, facilities, and the environment to prevent contamination of natural resources and to promote public health and safety.

Hazardous materials removal workers use a variety of tools and equipment, depending on the work at hand. Equipment ranges from brooms to personal protective suits that completely isolate workers from the hazardous material. Because of the threat of contamination, workers often wear disposable or reusable coveralls, gloves, hardhats, shoe covers, safety glasses or goggles, chemical-resistant clothing, face shields, and devices to protect one's hearing.

Most workers are also required to wear respirators while working to protect them from airborne particles or noxious gases. The respirators range from simple versions that cover only the mouth and the nose to self-contained suits with their own air supply. Recent improvements to respiratory equipment allow for greater comfort, enabling workers to wear the equipment for a longer period of time.

Asbestos and Lead

Asbestos and lead are two of the most common contaminants that hazardous materials removal workers encounter. Through the 1970s, asbestos was used to fireproof roofing and flooring, for heat insulation, and for a variety of other purposes. It was durable, fire retardant, corrosion resistant, and well insulated, making it ideal for such applications. Embedded in materials, asbestos is fairly harmless; airborne as a particulate, however, it can cause several deadly lung diseases, including lung cancer and asbestosis.

Today, asbestos is rarely used in buildings, but there are still structures that contain this material that must be remediated. Similarly, lead was a common building element found in paint and plumbing fixtures and pipes until the late 1970s. Because lead is easily absorbed into the bloodstream, often from breathing lead dust or from eating chips of paint containing lead, it can cause serious health risks, especially in children. Due to these risks, it has become necessary to remove lead-based products from buildings and structures.

Asbestos abatement workers and lead abatement workers remove asbestos, lead, and other materials from buildings scheduled to be renovated or demolished. Using a variety of hand and power tools, such as vacuums and scrapers, these workers remove the asbestos and lead from surfaces.

A typical residential lead abatement project involves the use of a chemical to strip the lead-based paint from the walls of the home. Lead abatement workers apply the compound with a putty knife and allow it to dry. Then they scrape the hazardous material into an impregnable container for transport and storage. They also use sandblasters and high-pressure water sprayers to remove lead from larger structures.

The vacuums utilized by asbestos abatement workers have special, highly efficient filters designed to trap the asbestos, which later is disposed of or stored. During the abatement, special monitors measure the amount of asbestos and lead in the air to protect the workers; in addition, lead abatement workers wear a personal air monitor that indicates the amount of lead to which a worker has been exposed. Workers also use monitoring devices to identify the asbestos, lead, and other materials that need to be removed from the surfaces of walls and structures.

The tables and figures in this chapter show examples of what is required of asbestos workers. The illustrations range from work requirements to paperwork. You can see them in Table 10.1 and Figures 10.1 through 10.5. They include both work forms and checklists.

▶ TRANSPORTATION

Transportation of hazardous materials is safer today than it was in the past, but accidents still occur. Emergency and disaster response workers clean up hazardous materials after train derailments and trucking accidents. These workers also are needed when an immediate cleanup is required, as would be the case after an attack by biological or chemical weapons.

Some hazardous materials removal workers specialize in radioactive substances. These substances range from low-level-contaminated protective clothing, tools, filters, and medical equipment to highly radioactive nuclear reactor fuels. Decontamination technicians

TABLE 10.1. Existing NESHAP Requirements Summary

	DEMOLITION		RENOVATION	
Amount* (in 1 yr.)	>260 ln. ft. or >160 sq. ft.	<260 ln. ft. or <160 sq. ft.	>260 ln. ft. or >160 sq. ft.	<260 ln. ft. or >160 sq. ft.
Notification	Yes	Yes	Yes	Not required
How far in advance*	10 Days	20 Days	As soon as possible	Not required
Emission controls (work practices)	Yes	Not required	Yes	Not required
Disposal standard	Yes	Not required	Yes	Not required

*May be changed on promulgation of Revised NESHAP Rule.

A sample application form for maintenance work approval
Job Request Form for Maintenance Work

Name:_____ Date:_____

Telephone No._____ Job Request No._____

Requested starting date:_____ Anticipated finish date:_____

Address, building, and room number(s) (or description of area) where work is to be performed:

_____
_____
_____
_____

Description of work:

_____
_____
_____
_____
_____

Description of any asbestos-containing material that might be affected, if known (include location and type):

_____
_____
_____
_____
_____

Name and telephone number of requestor:

_____

Name and telephone number of supervisor:

_____

Submit this application to

_____
(Asbestos Program Manager)

NOTE: An application must be submitted for all maintenance work whether or not asbestos-containing material might be affected. An authorization must then be received before any work can proceed.

_____ Granted (Job Request No. _____)
_____ With conditions*
_____ Denied

*Conditions _____
_____

Figure 10.1 Job Request Form for Maintenance Work.

Maintenance Work Authorization Form   No._____

AUTHORIZATION

Authorization is given to proceed with the following maintenance work:

_____

_____

_____

_____

_____

_____

_____

_____

PRESENCE OF ASBESTOS-CONTAINING MATERIALS

_____ Asbestos-containing materials are not present in the vicinity of the maintenance work.

_____ ACM is present, but its disturbance is not anticipated; however, if conditions change, the Asbestos Program Manager will reevaluate the work request prior to proceeding.

_____ ACM is present and may be disturbed.

Work Practices if Asbestos-Containing Materials Are Present

The following work practices shall be employed to avoid or minimum disturbing asbestos:*

_____

_____

_____

_____

_____

Personal Protection if Asbestos-Containing Materials Are Present**

The following equipment/clothes shall be used/worn during the work to protect workers:

_____

_____

_____

_____

(manuals on personal protection can be referenced)

Special Practices and/or Equipment Required:

_____

_____

Signed:_____ Date:_____
                    (Asbestos Program Manager)

Figure 10.2 Maintenance Work Authorization Form.

This evaluation covers the following maintenance work:

Location of work (address, building, room number(s), or general description):

_____

_____

_____

_____

Date(s) of work: _____

Description of work: _____

Work approval form number: _____

Evaluation of work practices employed to minimize disturbance of asbestos:

_____

_____

_____

_____

_____

Evaluation of work practices employed to contain released fibers and to clean up the work area:

_____

_____

_____

_____

_____

Evaluation of equipment and procedures used to protect workers:

_____

_____

_____

_____

_____

Personal air monitoring results (in-house worker or contract?)

Worker name_____ Results:_____

Worker name_____ Results:_____

Handling or storage of ACM waste: _____

Signed _____ Date: _____
                (Asbestos Program Manager)

Figure 10.2 *Cont'd*

Work Permit Application

1. Address, building, and room number (or description) where work is to be performed:

   _____

   _____

2. Requested starting date:_____ Anticipated finish date:_____

3. Description of work:_____

   _____

   _____

   _____

   _____

   _____

4. Description of any asbestos-containing material that might be affected, if known (include location and type):

   _____

   _____

   _____

   _____

   _____

   _____

   _____

5. Name and telephone number of requestor:_____

6. Name and telephone number of supervisor:_____

   Submit this application to the asbestos program manager.

   NOTE: An application must be submitted for all maintenance work whether or not asbestos-containing material might be affected. This authorization must then be signed before any work can proceed.

   _____ Granted (Work Permit No.____)

   _____ Denied (see Asbestos Program Manager)

   _____ Denied (until further sampling is conducted)

   Signed:_____ Date:_____
   (Asbestos Program Manager)

Figure 10.3 Work Permit Application.

Oklahoma Department of Labor
Mark Costello, Commissioner
www.labor.ok.gov

Oklahoma City
3017 N Stiles, Suite 100
Oklahoma City, OK 73105
405-521-6464
888-269-5353
Fax 405-521-6025

Tulsa
440 S Houston, Suite 300
Tulsa, OK 74127
918-581-2400
Fax 918-581-2431

ASBESTOS PROJECT CHECKLIST

☐ Initial Notification   ☐ Revised Notification   ☐ Emergency Notification

	NAME	ADDRESS	CITY	PHONE
Job Site:				
Contractor:				
Site Owner:				
Gen. Contractor:				
Project Designer:				
Air Monitoring Firm:				
Air Monitoring Firm:				
Landfill:				
Hauler:				

MOBILIZATION DATE: _____   SCHEDULED DATE OF ASBESTOS REMOVAL: _____

PROJECT COMPLETION DATE: _____   RENOVATION: ☐   DEMOLITION: ☐   EMERGENCY: ☐

Type and percentage asbestos (attach lab reports): _____

AMOUNT OF ASBESTOS TO BE ABATED: _____

ABATEMENT TECHNIQUES: _____

SUBMITTALS NECESSARY BEFORE ABATEMENT MAY BEGIN. CHECK OFF <u>ONLY</u> THOSE ATTACHED TO THIS CHECKLIST OR WHICH ARE ON FILE AT THE OKLAHOMA STATE DEPARTMENT OF LABOR.

☐ NESHAPS Notification (Copy)   Variances

☐ Project specifications

☐ Bonds and/or Insurance Certificates

☐ Plans for Decontamination Facilities

☐ Respirator Program

☐ Employee Physicals

☐ Permission from owner for all rented vehicles/trailers used to haul asbestos-containing material.

_____ # of Mini-containments   **FEES**
_____ # of Glove-bags   * $600.00 Per containment.
_____ # of Containments   * $200.00 Per project not part of a definite containment.
_____ # of Phases   * $200.00 Per project with multiple glove-bags or mini-containments, plus
                $5.00 per such glove-bag or mini-containment.

Comments: _____

_____

Revised: 01/10/2010   _____   _____
                        (Contractor/Responsible Party Signature)   (Date)

Figure 10.4 Asbestos Project Checklist.

**CHECKLIST FOR ASBESTOS REQUIREMENTS
FOR CONSTRUCTION PROJECTS**

BACKGROUND INFORMATION

Name of Auditor: _____

Date of Audit: _____

Name of Project/Site: _____

_____

A "notes" area is provided at the end of each section of this checklist. For every "No" answer, enter a description of the missing information and the action required to bring the site into compliance in the "notes" area.

ASBESTOS NESHAP

Yes	No	

Does the Asbestos NESHAP Apply to Site Activity?

The asbestos NESHAP will apply to the site activity if any of the following questions are answered "Yes." Only the notification requirements apply if the demolition activities contain Regulated Asbestos-Containing Material (RACM) below the following thresholds.

Yes	No	
❏	❏	1. Do the site renovations or demolitions include at least 80 linear meters (260 linear feet) of RACM on pipes?
❏	❏	2. Do the site renovations or demolitions include 15 square meters (160 square feet) of RACM on other facility components?
❏	❏	3. Do the site renovations or demolitions include at least 1 cubic meter (35 cubic feet) of facility components where the amount of RACM previously removed from pipes and other facility components could not be measured before stripping?
❏	❏	4. Do the site renovations or demolitions occur at residential structures with five or more dwellings (i.e., apartments or single-family homes)?

Will Category I ACM Become RACM and Require Removal?

Category I ACM will become RACM if the answer to any of the following questions is "Yes."

Yes	No	
❏	❏	5. Is Category I material friable or in poor condition?
❏	❏	6. Has the Category I material been or will it be subjected to sanding, cutting, grinding, or abrading?

Figure 10.5 Checklist for Asbestos Requirements for Construction Projects.

Yes	No	
❏	❏	7. Has a floor tile removal process, such as using a shot-blaster, resulted in extensive damage to the tiles?
❏	❏	8. Is debris from Category I roofing material created by sawing activities?
❏	❏	9. Will a building containing asbestos-cement products be demolished using cranes, hydraulic excavaters, or implosion/explosion techniques?
❏	❏	10. Will Jackhammers or other mechanical devices be used to break up asbestos-containing concrete or other materials coated with Category I nonfriable ACM?
❏	❏	11. Will bulldozers, tree chippers, or other equipment be used to reduce the volume of Category I materials?

Will Category II ACM Become RACM and Require Removal?

Category II ACM will become RACM if the answer to any of the following questions is "Yes."

❏	❏	12. Has Category II material been or will it be subjected to sanding, cutting, grinding, or abrading?
❏	❏	13. Will demolition activities be conducted using heavy equipment such as bulldozers and hydraulic excavaters?
❏	❏	14. Will equipment such as wrecking balls or buckets be used in demolishing asbestos cement?
❏	❏	15. Will the building be demolished using explosion/implosion?

Notification

The following notification requirements apply if the renovation/demolition activities have RACM below the thresholds listed above.

❏	❏	16. Did the site provide the Administrator with a complete written notice of intention to demolish or renovate? The notification must be postmarked no later than 10 working days before any striping/removal activity begins and must contain the following:
❏	❏	Indication if this is the original or a revised notification.
❏	❏	Name, address, and telephone number of both the facility owner and operator and the asbestos removal contractor owner or operator.
❏	❏	Type of operation: demolition or renovation.

Figure 10.5 *Cont'd*

Yes	No	
❑	❑	Description of the facility or affected part of the facility including the size (square feet and number of floors), age, and present and prior use of the facility.
❑	❑	The procedure, including analytical methods, used to detect the presence of RACM and Category I and Category II nonfriable ACM.
❑	❑	The approximate amount of RACM to be removed from the facility in terms of length of pipe in linear meters (linear feet), surface area in square meters (square feet) on other facility components, or volume in cubic meters (cubic feet) if off the facility components.
❑	❑	Approximate amount of Category I and Category II nonfriable ACM in the affected part of the facility that will not be removed before demolition.
❑	❑	Location and street address (including building number or name and floor or room number, if appropriate), city, county, and state of the facility being demolished or renovated.
❑	❑	Scheduled start and completion dates of asbestos removal work for the demolition or renovation.
❑	❑	Scheduled start and completion dates of demolition or renovation.
❑	❑	Description of work practices and engineering controls to be used to comply with the requirements of this subpart, including asbestos removal and waste-handling emission control procedures.
❑	❑	The name and location of the waste disposal site where the asbestos-containing waste material will be deposited.
❑	❑	Certification that at least one person trained as required by the asbestos NESHAP will supervise the stripping and removal described by the notification.
❑	❑	If the building is being demolished because it has been declared unsound, the name, title, and authority of the state or local government representative who has ordered the demolition, the date that the order was issued, the date on which the demolition was ordered to begin, and a copy of the order shall be attached to the notification.
❑	❑	For emergency renovations, the date and hour that the emergency occurred; a description of the sudden, unexpected event; and an explanation of how the event caused an unsafe condition, or would cause equipment damage or an unreasonable financial burden.

Figure 10.5 *Cont'd*

Yes	No	
❏	❏	Description of procedures to be followed in the event that unexpected RACM is found or Category II nonfriable ACM becomes crumbled, pulverized, or reduced to powder.
❏	❏	The name, address, and telephone number of the waste transporter.
❏	❏	17. If the amount of asbestos in the renovation/demolition changed by at least 20 percent, did the site update the notice?
❏	❏	18. If the actual start date is after the start date in the original notification, did the site provide the Administrator with a written notification with the new start date?
❏	❏	19. If the actual start date is before the start date in the original notification, did the site provide the Administrator with a written notification with the new start date and notify by telephone as soon as possible before the original start date?

NOTES/ACTIONS NEEDED TO BRING SITE INTO COMPLIANCE: _____

_____

_____

_____

_____

_____

_____

_____

Figure 10.5 *Cont'd*

ASBESTOS EMISSION CONTROL

If the answer is "Yes" to the following applicable questions, the site is complying with the procedures for asbestos emission control. No visible emissions are allowed under any circumstances.

Yes	No	
☐	☐	20. Has all RACM from the facility being demolished or renovated been removed before any activity begins that would break up, dislodge, or similarly disturb the material or preclude access to the material for subsequent removal? OR If the building is being demolished because it has been declared structurally unsound and in danger of imminent collapse or RACM is discovered after demolition, is the portion of the facility containing RACM kept adequately wet?
☐	☐	21. Is at least one on-site representative, such as a foreman or management-level person or other authorized representative, trained in the provisions of the asbestos NESHAP and the means of complying with them present during the demolition or renovation activities?
☐	☐	22. If the facility is being demolished by intentional burning, has all RACM including all Category I and Category II nonfriable ACM been removed?

When a facility component that contains, is covered with, or is coated with RACM is being taken out of the facility as a unit or in sections:

Yes	No	
☐	☐	23. Is the RACM that is exposed during cutting or disjoining operations adequately wet?
☐	☐	24. Does the site carefully lower each unit or section to the floor and to ground level to avoid damaging or disturbing the RACM?

When RACM is stripped from a facility component but it remains temporarily in place in the facility:

Yes	No	
☐	☐	25. Does the site adequately wet the RACM during the stripping operation? OR
		26. If the site does not wet the RACM:
☐	☐	Does the site obtain prior written approval from the Administrator? OR
☐	☐	Does the site use a local exhaust ventilation and collection system to capture the particulate asbestos material produced by the stripping? OR

Figure 10.5 *Cont'd*

Yes	No	
❏	❏	Does the site use a glove-bag system or use leak-tight wrapping to contain the particulate asbestos material? OR
❏	❏	Will the site use an alternate control method (besides a local exhaust ventilation and collection or a glove-bag system) and obtain written approval from the Administrator?

After a facility component covered with, coated with, or containing RACM has been taken out of the facility as a unit or in sections, it shall be stripped or contained in leak-tight wrapping.

❏	❏	27. If the site does not wet the RACM, does it use a local exhaust ventilation and collection system to capture the particulate asbestos material produced by the stripping?

For large facility components such as reactor vessels or large tanks, the RACM is not required to be stripped if the site meets the following requirements:

❏	❏	28. Is the component removed, transported, stored, disposed of, or reused without disturbing or damaging the RACM?
❏	❏	29. Is the component encased in a leak-tight wrapping and properly labeled?

For all RACM, including material that has been removed or stripped:

❏	❏	30. Is the material adequately wet and does it remain wet until collected and contained or treated in preparation for disposal?
❏	❏	31. Does the site carefully lower the material to the ground and floor, not dropping, throwing, sliding, or otherwise damaging or disturbing the material?
❏	❏	32. If the material has been removed or stripped more than 50 feet above ground level and was not removed as units or in sections, does the site transport the material to the ground via leak-tight chutes or containers?

When the temperature at the point of wetting is below 0°C (32°F):

❏	❏	33. Does the site remove facility components containing, coated with, or covered with RACM as units or in sections to the maximum extent possible?
❏	❏	34. During periods when wetting operations are suspended due to freezing temperatures, does the site record the temperature in the area containing the facility components at the beginning, middle, and end of each workday?

Figure 10.5 *Cont'd*

NOTES/ACTIONS NEEDED TO BRING SITE INTO COMPLIANCE: _____

_____

_____

_____

_____

_____

_____

_____

ASBESTOS WASTE DISPOSAL

Waste disposal requirements of the asbestos NESHAP do not apply to Category I and Category II nonfriable ACM waste that did not become crumbled, pulverized, or reduced to powder. If the answer to the following applicable questions below is "Yes," the site is complying with the procedures for asbestos waste disposal.

Yes	No	
❏	❏	35. Did the site properly mark vehicles used to transport asbestos-containing waste material so that the signs are visible?

To ensure no discharge of visible emissions to the outside air during the collection, processing (including incineration), packaging, or transporting of any asbestos-containing waste material generated by the source, the site must use all applicable emission control and waste treatment methods.

❏	❏	36. Does the site adequately wet asbestos-containing waste material?
❏	❏	37. Does the site seal all asbestos-containing waste material in leak-tight containers while wet or put materials into leak-tight wrapping?
❏	❏	38. Does the site properly label the containers or wrapped materials?
❏	❏	39. Does the label contain the name of the waste generator and the location at which the waste was generated?
❏	❏	40. Does the site process asbestos-containing waste material into nonfriable forms?

Figure 10.5 *Cont'd*

Yes	No	
☐	☐	41. If the building is being demolished because it has been declared unsound:
☐	☐	Does the site adequately wet asbestos-containing waste material at all times after demolition and keep wet during handling and loading for transport to a disposal site?
		OR
☐	☐	Does the site use an alternative emission control and waste treatment method that has received prior approval by the Administrator?

All asbestos-containing waste material shall be disposed as soon as is practical by the waste generator. This does not apply to Category I and Category II nonfriable ACM that is not RACM.

Yes	No	
☐	☐	42. Is the material sent to an active waste disposal site authorized to receive asbestos-containing material?
☐	☐	43. Is the material sent to an EPA-approved site that converts asbestos-containing waste material into asbestos-free material?

For all asbestos-containing waste material transported off the facility site:

Yes	No	
		44. Does the site maintain waste shipment records using the required format?
☐	☐	Name, address, and telephone number of the waste generator.
☐	☐	Name and address of the local, state, or EPA regional office responsible for administering the asbestos NESHAP program.
☐	☐	Approximate quantity of waste in cubic meters.
☐	☐	Name and telephone number of the disposal site operator.
☐	☐	Name and physical site location of the disposal site.
☐	☐	Date the shipment was transported.
☐	☐	Name, address, and telephone number of the transporter.
☐	☐	Certification that the contents of this consignment are accurately described by proper shipping name and are in proper condition for transport by highway.

Figure 10.5 *Cont'd*

Yes	No	
☐	☐	45. Did the site provide a copy of the waste shipment record to the disposal site owners or operators as the asbestos-containing waste material was delivered to the disposal site?
☐	☐	46. Has the site retained a copy of all waste shipment records for at least two years?
☐	☐	47. If the site did not receive a copy of the waste shipment record signed by the owner or operator of the designated disposal site within 45 days of the date the waste was accepted by the initial transporter, was the transporter and/or the owner or operator of the designated disposal site contacted to determine the status of the waste shipment?
☐	☐	48. If a copy of the waste shipment record, signed by the owner or operator of the designated waste disposal site, was not received within 45 days of the date the waste was accepted by the initial transporter, did the site report this in writing to the office responsible for administering the asbestos NESHAP program for waste generators?

NOTES/ACTIONS NEEDED TO BRING SITE INTO COMPLIANCE: _____

_____

_____

_____

_____

_____

_____

_____

Figure 10.5 *Cont'd*

AIR CLEANING

If the site chooses to use a local exhaust ventilation and collection system to capture particulate asbestos material emissions, it must be designed and operated in accordance with the requirements of 40 CFR, Part 61, Subpart 152. The site is complying with this subpart if the answer to all of the following applicable questions is "Yes."

Yes   No

If the site uses fabric filter devices:

❏	❏	49. Does the site ensure that the airflow permeability does not exceed 9 m³/min/m² (30 ft³/min/ft²) for woven fabrics or 11 m² (35 ft³/min/ft²) for felted fabrics?
❏	❏	50. Does the site ensure that felted fabric weighs at least 475 grams per square meter (14 ounces per square yard) and is at least 1.6 millimeters (one-sixteenth inch) thick throughout?
❏	❏	51. Does the site avoid the use of synthetic fabrics that contain fill yarn other than that which is spun?
❏	❏	52. Does the site properly install, use, operate, and maintain all air-cleaning equipment?

If the site does not use fabric filter devices:

❏	❏	53. Does the site utilize wet collectors designed to operate with a unit contacting energy of at least 9.95 kilopascals (40 inches water gage pressure)? OR
❏	❏	Does the site use a HEPA filter that is certified to be at least 99.97 percent efficient for 0.3 micron particles? OR
❏	❏	Is the site authorized to use alternative filtering equipment?

NOTES/ACTIONS NEEDED TO BRING SITE INTO COMPLIANCE:_____

_____

_____

_____

_____

_____

Figure 10.5 *Cont'd*

perform duties similar to those of janitors and cleaners, but the items and areas they clean are radioactive. They use brooms, mops, and other tools to clean exposed areas and remove exposed items for decontamination or disposal. Some of these jobs are now being done by robots controlled by people away from the contamination site. Increasingly, many of these remote devices are being used to automatically monitor and survey surfaces, such as floors and walls, for contamination.

With experience, decontamination technicians can advance to radiation-protection technician jobs and use radiation survey meters and other remote devices to locate and assess radiated materials, operate high-pressure cleaning equipment for decontamination, and package radioactive materials for transportation or disposal.

Decommissioning and decontamination workers remove and treat radioactive materials generated by nuclear facilities and power plants. With a variety of hand tools, they break down contaminated items such as glove boxes, which are used to process radioactive materials. At decommissioning sites, the workers clean and decontaminate the facility, as well as remove any radioactive or contaminated materials.

Treatment, storage, and disposal workers transport and prepare materials for treatment or disposal. To ensure proper treatment of materials, laws enforced by the U.S. Environmental Protection Agency (EPA) or Occupational Safety and Health Administration (OSHA) require these workers to be able to verify shipping manifests. At incinerator facilities, treatment, storage, and disposal workers transport materials from the customer or service center to the incinerator. At landfills, they follow a strict procedure for the processing and storage of hazardous materials. They organize and track the location of items in the landfill and may help change the state of a material from liquid to solid in preparation for its storage. These workers typically operate heavy machinery, such as forklifts, earth-moving machinery, and large trucks and rigs.

▶ THE SUPERFUND

To help clean up the nation's hazardous waste sites, a federal program called Superfund was created in 1980. Under the Superfund program, abandoned, accidentally spilled, or illegally dumped hazardous waste that poses a current or future threat to human health or the environment is cleaned up. In doing so, the EPA, along with potentially responsible parties, communities, local, state, and federal authorities, identify hazardous waste sites, test site conditions, devise cleanup plans, and clean up the sites.

▶ MOLD

Mold remediation is a new aspect of some hazardous materials removal work. Some types of mold can cause harsh allergic reactions, especially in people who are susceptible to them. Although mold is present in almost all structures and is not usually defined as a hazardous material, some mold—especially the types that cause allergic reactions—can infest a building to such a degree that extensive efforts must be taken to remove it safely.

Molds are fungi that typically grow in warm, damp conditions both indoors and outdoors yearround. They can be found in heating and air conditioning ducts; within walls; and in showers, attics, and basements. Although mold remediation is often undertaken by other construction workers, large-scale mold removal is usually handled by hazardous materials removal workers, who take special precautions to protect themselves and surrounding areas from being contaminated.

▶ SUPERVISION

Hazardous materials removal workers may also be required to construct scaffolding or erect containment areas prior to abatement or decontamination. In most cases, government regulation dictates that hazardous materials removal workers be closely supervised on the work site. The standard usually is 1 supervisor to every 10 workers. The work is highly structured, sometimes planned years in advance, and usually team oriented. There is a great deal of cooperation among supervisors and workers. Because of the hazard presented by the materials being removed, work areas are restricted to licensed hazardous materials removal workers, thus minimizing exposure to the public.

▶ WORK ENVIRONMENT

Hazardous materials removal workers function in a highly structured environment to minimize the danger they face. Each phase of an operation is planned in advance, and workers are trained to deal with hazardous situations. Crews and supervisors take every safety measure to ensure that the work site is safe. Whether they work with asbestos, mold, lead abatement, or in radioactive decontamination, hazardous materials removal workers must stand, stoop, and kneel for long periods. Some must wear fully enclosed personal protective suits for several hours at a time; these suits may be hot and uncomfortable and may cause some individuals to experience claustrophobia.

Hazardous materials removal workers face different working conditions, depending on their area of expertise. Although many work a standard 40-hour week, overtime and shift work are common, especially for emergency and disaster response workers. Asbestos and lead abatement workers usually work in structures such as office buildings, schools, or historic buildings under renovation. Because they are under pressure to complete their work within certain deadlines, workers may experience fatigue. Completing projects frequently requires night and weekend work, because hazardous materials removal workers often work around the schedules of others.

Treatment, storage, and disposal workers are employed primarily at facilities such as landfills, incinerators, boilers, and industrial furnaces. These facilities often are located in remote areas, due to the kinds of work being done, so workers may have to commute long distances to their jobs.

Decommissioning and decontamination workers, decontamination technicians, and radiation protection technicians work at nuclear facilities and electric power plants. Like treatment, storage, and disposal facilities, these sites are often far from urban areas. Workers who perform jobs in cramped conditions may need to use sharp tools to dismantle contaminated objects. A hazardous materials removal worker must have great self-control and a level head to cope with the daily stress associated with handling hazardous materials.

Hazardous materials removal workers may be required to travel outside their normal working areas to respond to emergency cleanups, which sometimes take several days or weeks to complete. During the cleanup, workers may be away from home for the entire time.

▶ TRAINING, OTHER QUALIFICATIONS, AND ADVANCEMENT

No formal education beyond a high school diploma is required for a person to become a hazardous materials removal worker. However, federal, state, and local government standards require

---

Box 10.1. Key Points about Training and Accreditation

AHERA does not require that **designated persons** complete EPA- or state-approved courses and become accredited, but § 763.84(g)(2) of the AHERA Rule requires that training for the designated persons provide basic knowledge of a number of asbestos-related subjects.

The LEA must ensure that all maintenance and custodial staff who may work in a building that contain ACBM receive a minimum of **2 hours awareness training**, whether or not they are required to work with ACBM. All new maintenance and custodial staff must receive asbestos awareness training within 60 days of hire.

Staff who may disturb ACBM must receive an additional **14 hours** of training.

Building inspectors, management planners, project designers, contractors/supervisors, and asbestos workers must successfully complete EPA- or state-approved courses, pass an exam, and **receive accreditation** before they can perform any asbestors-related activities.

Building inspectors, management planners, project designers, contractors/supervisors, and asbestos workers must complete annual EPA- or state-approved **refresher courses** to maintain their accreditation.

---

specific types of on-the-job training. Regulations vary by specialty and sometimes by state or locality. Employers are responsible for employee training. Some of the key points for this duty are shown in Box 10.1.

Hazardous materials removal workers usually need at least 40 hours of formal on-the-job training. For most specialties, this training must meet specific requirements set by the federal government or individual states.

Workers who treat asbestos and lead, the most common contaminants, must complete a training program through their employer that meets OSHA standards. Employer-sponsored training is usually performed in-house, and the employer is responsible for covering all technical and safety subjects outlined by OSHA. Accredited personal training requirements are shown in Table 10.2.

To become an emergency and disaster response worker and treatment, storage, and disposal worker, a candidate must obtain a federal license as mandated by OSHA. Employers are responsible for ensuring that employees complete a

TABLE 10.2. Accredited Personnel Training Requirements

JOB TITLE	SUBJECT MATTER OF TRAINING	AMOUNT OF TRAINING (DAYS)	ANNUAL TRAINING UPDATE (DAYS)
Building Inspectors	Technical information needed to identify and describe ACBM; information needed to write an inspection report	3	1/2
Management Planners	Extension of the building inspector training, plus how to develop a schedule (or plan) for implementation of response actions for hazards or potential hazards identified in the inspection report, how to develop an O&M plan, and how to prepare a management plan	2[a]	1[b]
Project Designers	How to design response actions and abatement projects; basic concepts of architectural design, engineering controls and proper work practices	3	1
Contractors/Supervisors	Proper work practices and procedures; contractor issues such as legal liability, contract specifications, insurance, and bonding; air monitoring	5	1
Asbestos Workers	Work practices and procedures, personal protective equipment, health effects of asbestos exposure, and other critical information	4	1

[a]Management planners must first complete the building inspector training and pass the exam.
[b]This includes the half-day building inspector training update.

formal 40-hour training program, given either in-house or in OSHA-approved training centers. The program covers health hazards, personal protective equipment and clothing, site safety, recognition and identification of hazards, and decontamination.

In some cases, workers may discover one hazardous material while abating another. If workers are not licensed to handle the newly discovered material, they cannot continue to work with it. See Table 10.3 for training requirements. Many experienced workers opt to take courses in additional types of hazardous materials removal to avoid this situation.

Mold removal is not regulated by OSHA, but is regulated by each state. For decommissioning and decontamination workers employed at nuclear facilities, training is most extensive. In addition to obtaining licensure through the standard 40-hour training course in hazardous waste removal, workers must take courses dealing with regulations governing nuclear materials and radiation safety as mandated by the Nuclear Regulatory Commission. These courses add up to approximately 3 months of training, although most are not taken consecutively.

Many agencies, organizations, and companies throughout the country provide training programs that are approved by the U.S. Environmental Protection Agency, the U.S. Department of Energy, and other regulatory bodies. To maintain their license, workers in all fields are required to take continuing-education courses as a refresher, every year.

Workers must be able to perform basic mathematical conversions and calculations when mixing solutions that neutralize contaminants and should have good physical strength and manual dexterity. Because of the nature of the work and the time constraints sometimes involved, employers prefer people who are dependable, prompt, and detail oriented. Because much of the work is done in buildings, a background in construction is helpful. This makes the work a natural enhancement for building or remodeling contractors to add to their credentials. Then you can have more control over your own jobs. Examples of various license applications are provided in Figures 10.6 through 10.8.

TABLE 10.3. LEA Employee Training Requirements

JOB TITLE	SUBJECT MATTER OF TRAINING	AMOUNT OF TRAINING (HOURS)	ANNUAL TRAINING UPDATE (HOURS)
Designated Person	Health effects of asbestos; detection, identification, and assessment of ACBM; options for controlling ACBM; asbestos management program; related federal and state laws	Adequate	None
All Maintenance Workers	Asbestos and its uses and forms; health effects associated with asbestos exposure; locating ACBM identified throughout each school building in which they work; recognizing various conditions of ACBM; name and telephone number of LEA designated person; information pertaining to the availability and location of management plan	2	None
Maintenance Workers Who Disturb ACBM	Proper methods for handling ACBM; information on proper use of respiratory protection; hands-on training in the use of respiratory protection, other personal protection measures, and good work practices; information pertaining to various regulations; technical information	16 (asbestos awareness and 14 additional hours*)	None

*These 14 hours of training are in addition to the 2 hours of asbestos awareness training that all maintenance workers receive. Note that state and local requirements may be more stringent.

Oklahoma Department of Labor

Mark Costello
COMMISSIONER OF LABOR

ASBESTOS ABATEMENT
WORKER APPLICATION

☐ NEW   ☐ RENEWAL   LICENSE #

1. Applicant's Name			
2. Home Address			
3. City	4. State	5. Zip	
6. Date of Birth	7. Social Security Number	8. Phone	
8. Hair Color	9. Eye Color	10. Weight	11. Height
12. Company Name			
13. Company Address			
14. City	15. State	16. Zip	
17. Company Phone		18. Company Contact Person	

FOR OFFICE USE ONLY				
Date	DEO	Receipt No.	License No.	Asbestos Admin Approval/Date
Type of Payment	1 2 3	#	Endorser	

Figure 10.6 Oklahoma's four-page Asbestos Abatement Worker Application for renewal of license.

Do you currently hold any other Oklahoma Asbestos Licenses? ☐ YES   ☐ NO

If yes, please indicate the type of license and the license number:

CHECK TYPE LICENSE	LICENSE NUMBER
Supervisor	
Mgmt. Planner	

CHECK TYPE LICENSE	LICENSE NUMBER
Inspector	
Project Designer	

Please check one of the following:

	NON-EXEMPT	Submit check or money order for $25.00 made payable to the Oklahoma Department of Labor
	OR	
	EXEMPT	If the applicant is the employee of a political subdivision of the State of Oklahoma and agrees to perform any AHERA asbestos work on behalf of that employer only, no fee is required.

ALL APPLICANTS MUST SUBMIT THE FOLLOWING:

1.   Copy(s) of current refresher training class*.
2.   Provide current drivers license or state issued photo identification card.
3.   Provide social security card or letter from social security administration showing you have reapplied for your card.

NEW APPLICANTS MUST ALSO SUBMIT THE FOLLOWING, IN PERSON:

1.   Copies of Original Worker or Contractor/Supervisor training class and all subsequent refresher courses.*
2.   Provide current drivers license or state issued photo identification card.
3.   Provide social security card or letter from social security administration showing you have reapplied for your card.
4.   Affidavit regarding citizenship.

*Training must have been provided by an educational institution, government agency or labor union and must have been accredited by the U.S. Environmental Protection Agency.

Do you have any asbestos violations?   ☐ YES   ☐ NO

If yes, please list job and type of violation: _____

_____

_____

_____

Figure 10.6 *Cont'd*

I hereby authorize the educational institutions to release verification of completion of the courses presented in this application.

I further affirm, upon my oath, to follow Title 40 of the Oklahoma Statutes, Section 450 through 456 as amended, and the <u>Abatement of Friable Asbestos Materials Rules OAC 380, Chapter 50.</u> I also attest that my license is valid only when working for an Oklahoma licensed Contractor. I understand that a violation of any law or rule may subject my license to be suspended or revoked, or may subject me to cease and desist orders, injunctive measures, and criminal penalties for criminal violations.

I, upon my oath, do state that the above information is a true statement, and further state that I am not under any type of disciplinary action, including license revocation or suspension, by any State or political division thereof, or by the United States government, for any illegal or improper activity, civil or criminal, involving asbestos-containing material.

_____   _____
Applicant's Signature   Date

NOTE: This application must be notarized

ACKNOWLEDGED

State of _____

County _____

Signed or affirmed before me this _____ day of _____, _____ .

_____
　　　　　Notary Public

My Commission Expires _____

Figure 10.6 *Cont'd*

OAC 380:50-5-9. Licensing of Asbestos Abatement Workers:

Licensing requirements for Asbestos Abatement Workers are as follows: (1) Applications shall be submitted on forms prescribed by the Commissioner. (2) The license fee shall be twenty-five dollars ($25.00) per year. (3) The license shall be issued for a period not to exceed one year and shall expire concurrently with the asbestos training and subsequent refresher training. There will be no grace period wherein a Worker will be allowed to work with an expired license. (4) Any Worker who has not taken the required refresher course within two years of the previous Asbestos Worker training or refresher course shall repeat the Asbestos Worker training requirements of OAC **380:50-6-2 and 380:50-6-7**. (5) Asbestos Abatement Workers shall have successfully completed and shall provide documentation for an Asbestos Abatement Worker's course and all subsequent Worker refresher training that fully meets the requirements of OAC **380:50-6-2 and 380:50-6-7**. (6) The licenses shall be issued in the name of the individual applicant and shall be valid only when working for a licensed Contractor. (7) License cards shall be available at the job site for inspection by the Department of Labor.

OAC 380:50-6-2. Initial training for Asbestos Workers:

Training requirements are as follows: (1) Training shall be supplied by an EPA or DOL accredited educational institution, labor union, or government agency, or from a private vocational education provider licensed by the state where it operates (pursuant to 70 O.S. § 21-103 within the state of Oklahoma and accredited by EPA or an EPA approved governmental agency. (2) Training providers must have EPA approval or approval by a state that has an approved Model Accreditation Plan (MAP) that is as stringent or exceeds the minimum requirements of the EPA MAP. (3) Asbestos Abatement Worker training courses shall be a minimum of four (4) days of instruction, including lectures, demonstrations, at least 14 hours of hands-on training, individual respirator fit testing, course review and an examination. (4) Asbestos Abatement Worker training courses shall consist of the training required in OAC **380:50-6-2 and 380:50-6-7**.

3017 N. STILES, OKLAHOMA CITY, OKLAHOMA 73105 • www.labor.ok.gov • PHONE: (405)521-6464 • FAX (405) 521-6016
Toll Free Number (888) 269-5353

Revised 2011-Jan

Page 4

Figure 10.6 *Cont'd*

Oklahoma Department of Labor

Mark Costello
COMMISSIONER OF LABOR

ASBESTOS ABATEMENT CONTRACTOR APPLICATION

☐ NEW   ☐ RENEWAL   LICENSE #

1. Applicant's Name- 1st Responsible Party			
2. Doing Business As		3. Social Security Number	4. Date of Birth
3. Address	4. City	5. State	6. Zip
7. Phone		8. FAX	

List the name and address of the Service Agent:

1. Name		2. Title & Capacity	
3. Address	4. City	5. State	6. Zip

List the names and home addresses of the officers, principals, partners and proprietors.

1. Name		2. Title & Capacity	
3. Address	4. City	5. State	6. Zip
1. Name		2. Title & Capacity	
3. Address	4. City	5. State	6. Zip
1. Name		2. Title & Capacity	
3. Address	4. City	5. State	6. Zip
1. Name		2. Title & Capacity	
3. Address	4. City	5. State	6. Zip

2nd Responsible Party (If applicable)

Name of Responsible Party: _____

Please indicate all current Oklahoma asbestos licenses and the license number:

CHECK TYPE LICENSE	LICENSE NUMBER		CHECK TYPE LICENSE	LICENSE NUMBER
Worker			Inspector	
Supervisor			Project Designer	
Mgmt. Planner				

FOR OFFICE USE ONLY				
Date	DEO	Receipt No.	License No.	Asbestos Admin Approval/Date
Type of Payment	1 2 3	#	Endorser	

3017 N. STILES, OKLAHOMA CITY, OKLAHOMA 73105 • www.labor.ok.gov • PHONE: (405)521-6464 • FAX (405) 521-6016
Revised 2011-Jan   Toll Free (888) 269-5353   Page 1

Figure 10.7 Oklahoma's four-page Asbestos Abatement Contractor Application for license renewal.

PLEASE SUBMIT COPIES OF THE FOLLOWING:

1.   Original Contractor/Supervisor training class and all subsequent refresher classes.*
2.   Other asbestos training classes*
3.   Proof of worker's compensation insurance coverage or a notarized statement explaining why such coverage is not mandated by law.

*Training must have been provided by an educational institution, government agency or labor union and must have been accredited by the U.S. Environmental Protection Agency.

I hereby authorize the educational institution(s) to release verification of completion of the courses presented in this application. I, upon my oath, do state that the above information is a true statement, and further state that I am not under any type of disciplinary action, including license revocation or suspension, by any State or political division thereof, or by the United States government, for any illegal or improper activity, civil or criminal, involving asbestos-containing material.

_____   _____
        Applicant's Signature   Date

Acknowledged

County_____

State_____

        Signed or affirmed before me this_____ day of_____,_____.

_____

Notary Public

My Commission expires:_____

Please list all locations in all states in which the Contractor has performed abatement of friable asbestos materials. Include the name and address of the local asbestos procedure enforcement agency(s) for each location.

1. Name of contracting authority:			
Address of contracting authority:			
Name of enforcing agency:			
Address of enforcing agency:	City	State	Zip
Project name:			
Project address:	City	State	Zip
2. Name of contracting authority:			
Address of contracting authority:			
Name of enforcing agency:			
Address of enforcing agency:	City	State	Zip
Project name:			
Project address:	City	State	Zip

Figure 10.7 *Cont'd*

3. Name of contracting authority:			
Address of contracting authority:			
Name of enforcing agency:			
Address of enforcing agency:	City	State	Zip
Project name:			
Project address:	City	State	Zip
4. Name of contracting authority:			
Address of contracting authority:			
Name of enforcing agency:			
Address of enforcing agency:	City	State	Zip
Project name:			
Project address:	City	State	Zip

Please submit the following with this application:

1. Approved Respirator Program
2. Operations and Maintenance Program

Please check one of the following:

	NON-EXEMPT	**New applicants only** must submit check or money order for $1,000.00 made payable to the Oklahoma Department of Labor as a nonrefundable application fee.

OR

	EXEMPT	The state and political subdivisions thereof shall be exempt from any certification fees when such entities act as a contractor.

OAC 380:50-5-5 of the Abatement of Friable Asbestos Materials Rules states: Licensing of Asbestos Abatement Contractors are as follows: (1) Applications shall be submitted on forms prescribed by the Commissioner. Submission of such application shall include a nonrefundable one thousand dollar ($1,000.00) processing fee. (2) After the statutory one hundred twenty (120) day waiting period, if a Contractor's application is accepted, the Contractor will be notified by the Commissioner and required to submit at that time the five hundred dollar ($500.00) license fee. (3) The applicant shall designate a minimum of one, or a maximum of two, responsible parties to be named on the license. Such responsible parties shall have and maintain the training credentials required for licensing. Documentation of satisfactory completion of the required training and all subsequent refresher training shall accompany the application. (A) In the absence of such responsible party in the employee of the Contractor, the Contractor will not be allowed to perform asbestos abatement work in the State of Oklahoma. (B) The responsible party shall have successfully completed and shall have documentation provided for not fewer than two asbestos training courses. One such course shall be an Asbestos Abatement Supervisor's course that fully meets the requirements of **OAC 380:50-6-3 and 380:50-6-8.** The Commissioner shall maintain updated lists of additional training courses acceptable for licensing. (C) Responsible parties may be changed or added to the license at any time, by paying a fee of fifty dollars ($50.00) per change or addition. Documentation of satisfactory completion of required training and all applicable subsequent refresher training shall be submitted. (4) Prior to issuance of the license, the Contractor must have a respirator program meeting all requirements of OSHA or DOL, whichever is most stringent. (5) Licenses shall be issued for a period of one year. (6) No Contractor may perform any asbestos abatement after expiration of the license. (7) If a Contractor allows the license to lapse for more than thirty (30) days, the license may not be renewed, and licensing will be permitted only after meeting all requirements for a new license, including the one hundred twenty (120) day waiting period. (8) License applicants must be of good character. Conviction for a felony by an applicant, if a proprietor or partner; by an officer, if a corporation; or by a responsible party, shall be grounds for denial of, or revocation of, a Contractor's license. (9) The Commissioner may refuse to issue an Asbestos Abatement Contractor's license to any applicant, if there are records of Notice of Violation (NOV) of NESHAP regulations by the applicant, or any principal, partner, or officer of the applicant's firm or associated firms, as maintained by EPA.

Figure 10.7 *Cont'd*

In accordance with Title 40, Section 452, Oklahoma Statutes as revised, there is a mandatory one-hundred twenty (120) day waiting period to allow the Asbestos Division to investigate the work record of the Applicant. The license may be denied to an applicant whose past performance for abatement of friable asbestos does not comply with Federal and other State requirements.

Providing false or incomplete information on this Affidavit/Application will be grounds for rejection of this application, or for suspension or revocation of the license, if issued.

At such time as the application and respiratory program are approved, the Applicant will be so notified. The Applicant at that time will submit the five-hundred-dollar ($500) annual license fee, and the license will be subsequently issued.

Foreign corporations must be registered with the Oklahoma Secretary of State.

Any supplemental information submitted with this application must be signed by the Applicant and notarized.

All agencies listed in this form may be contacted by the Oklahoma State Department of Labor to verify the information provided. Additionally, project-specific information may be requested from the applicant, regulatory agencies, or project owners during the licensing investigation.

I, upon my oath, do state that the above information is a true and correct statement.

I further affirm, upon my oath, to follow Title 40 of the Oklahoma Statutes, Sections 450 through 456 as amended, and any rules adopted by the Commissioner of Labor relative to any procedures and standards adopted thereto, and agree to abide by all Child Labor Laws, Federal and State, and any and all Workers Compensation Insurance Laws of the State of Oklahoma. I swear that none of the persons listed above have unpardoned felony convictions in any State or Federal court. I understand that a violation of any law or rule may subject my license to be suspended or revoked, or subject me to cease and desist orders, injunctive measures, and criminal penalties for willful violation.

_____   _____
Applicant's Signature   Date

Acknowledged

Country_____
State_____

        Signed or affirmed before me this_____day of_____,_____.

_____
Notary Public
My Commission expires:_____

_____   _____
Authorizing Signature & Seal (if corporation)   Date

Acknowledged

County_____
State_____

        Signed or affirmed before me this_____day of_____,_____.

_____
Notary Public
My Commission expires:_____

3017 N. STILES., OKLAHOMA CITY, OKLAHOMA 73105 • www.labor.ok.gov • PHONE: (405)521-6464 • FAX (405) 521-6016
Toll Free (888) 269-5353

Revised 2011-Jan   Page 4

Figure 10.7 *Cont'd*

Oklahoma Department of Labor

Mark Costello
COMMISSIONER

ELIGIBILITY APPLICATION

PLEASE CHECK ALL BOXES THAT APPLY

Are you an **Oklahoma Resident?** YES_____ NO_____

If yes, please provide the following two (2) forms of identification:

❑   A valid, unexpired **Oklahoma Driver's License** OR **Photo-Identification Card** from the Oklahoma Deartment of Public Safety; **AND**

❑   **Social Security Card**

If no, and you are **not** a **Resident of Oklahoma,** but a resident of another U.S. state, please provide the following two (2) forms of identification:

❑   A valid, unexpired **Driver's License** OR **Photo-Identification Card** from your state of residency; **AND**

❑   **Birth Certificate;** OR
Social Security Card; OR
Valid, unexpired **Passport;** OR
W-2 form from current employer

Are you a **citizen of the United States of America**? YES_____   NO_____

If yes, please provide the following document:

❑   **Affidavit of Citizenship** (See attached Affidavit on page 2)

Are you a **qualified alien lawfully present in the Untied States?** YES_____   NO_____

If yes, please provide the following documents:

❑   **Affidavit Regarding Citizenship** (see attached Affidavit on page 3)

❑   **ALL Qaulified Aliens** (non-U.S. Citizens) **must present** in person documentation to verify lawful presence. (See attached Affidavit on page 3)

Figure 10.8 Oklahoma's Eligibility Application for obtaining a license.

AFFIDAVIT OF CITIZENSHIP

I,_____, swear under penalty of perjury that I am a United States citizen.

I,_____, authorize the United States Department of Homeland Security to release my citizenship and immigration status to the Oklahoma Department of Labor in order to be eligible to receive the benefit/license for which I am applying.*

_____
Applicant's Signature

* Any person who knowingly and willfully makes a false, fictitious, or fraudulent statement of representation in this affidavit shall be subject to criminal penalties applicable in the State of Oklahoma for fraudulently obtaining a public assistance program benefit (a license). If the affidavit constitutes a false claim of U.S. citizenship under 18 U.S.C. Section 911, a complaint shall be filed by the Oklahoma Department of Labor with the United States Attorney General for the applicable district based on the venue in which the affidavit was executed.

ACKNOWLEDGMENT

State of Oklahoma   )
) ss:
County of_____)

    Subscribed and sworn to before me this_____ day of _____, 201___.

_____
Notary Public

My Commission expires: _____

Figure 10.8 *Cont'd*

ALL Qualified Aliens (non-U.S. Citizens)
must present in person one (1) of the following documents to verify lawful presence:
PLEASE CHECK BOX THAT APPLIES

☐ Unexpired foreign passport, with I-551 stamp, or attached Form I-94 indicating unexpired employment authorization
☐ Permanent Resident Card or Alien Registration Receipt Card with photograph (Form I-151, or I-551)
☐ Unexpired Temporary Resident Card (Form I-688)
☐ Unexpired Employment Authorization Card (Form I-688A)
☐ Unexpired Reentry Permit (Form I-327)
☐ Unexpired Refugee Travel Document (Form I-571)
☐ Unexpired Employment Authorization Document issued by DHS containing a photograph (Form I-688B)
☐ Valid unexpired immigrant or non-immigrant visa status for admission into the United States
☐ Pending or approved application for asylum in the United States
☐ Pending or approved application for temporary protected status in the United States
☐ Approved deferred action status (aliens whose deportation is being withheld under (1) §243(h) of the INA as in effect prior to April 1, 1997 or (2) §241(b)(3) of the INA)
☐ Pending application for adjustment of status to legal permanent resident or conditional resident status (aliens granted conditional entry under §203a)(7) of the INA before April 1, 1980). (Upon approval, the applicant may be issued a temporary license for the period of time of the authorized stay in the U.S., or if there is no definite end to the period of authorized stay, then for a period of one (1) year);
☐ Cuban and Haitian Entrants, as defined in §501(e) of the Refugee Education Assistance Act of 1980; aliens granted parole for at least one year under §212(d)(5) of the INA
☐ Battered aliens, who meet the conditions set forth in §431(c) of PRWORA, as added by §501 of the Illegal Immigration Reform and Immigrant Responsibility Act of 1996, P.L. 104-208 (IIRIRA), and amended by §5571of the Balanced Budget Act of 1997, P.L. 105-33 (BBA), and §1508 of the Violence against Women Act of 2000, P.L. 106-386; Section 431(c) of PRWORA, as amended, is codified at 8 U.S.C. 1641(c)
☐ Victims of a severe form of trafficking, in accordance with §107(b)(l) of the Trafficking Victims Protection Act of 2000, P.L. 106-386

AFFIDAVIT REGARDING CITIZENSHIP

I, _____, swear under a penalty of perjury, that I am a qualified alien lawfully present in the United States.

I, _____, authorize the United States Department of Homeland Security to release my citizenship and immigration status to the Oklahoma Department of Labor in order to be eligible to receive the benefit/license for which I am applying. *

Status: _____ Number:_____

_____
License Applicant

* Any person who knowingly and willfully makes a false, fictitious, or fraudulent statement of representation in this affidavit shall be subject to criminal penalties applicable in the State of Oklahoma for fraudulently obtaining a public assistance program benefit (a license). If the affidavit constitutes a false claim of U.S. citizenship under 18 U.S.C. Section 911, a complaint shall be filed by the Oklahoma Department of Labor with the United States Attorney General for the applicable district based on the venue in which the affidavit was executed.

ACKNOWLEDGMENT

State of Oklahoma   )
) ss:
County of_____)

Subscribed and sworn to before me this_____ day of _____, 201___.

_____   My Commission expires: _____
Notary Public

Figure 10.8 *Cont'd*

▶ HOW MUCH DO YOU ALREADY KNOW?

How much do you already know about working with hazardous materials? Would you like to find out? Take the following quiz and see just how much you know. The correct answers are at the end of the quiz.

1. Asbestos that is easily crumbled into a powder by hand pressure when dry is
   A. friable.
   B. nonfriable.

   C. decomposable.

   D. asbestos powder.

   E. none of the above.

2. Exposure to asbestos may result in

   A. asbestosis (a disease characterized by lung scarring).

   B. lung cancer.

   C. mesothelioma (a cancer arising in the chest cavity or abdominal cavity).

   D. all of the above.

   E. none of the above.

3. Asbestos-related diseases are _____ and have a latency period of _____.

   A. dosed related, 15 to 30 years

   B. fatal, 30 days

   C. nonexistent, 60 years

   D. dangerous, 1 hour

   E. serious, 10 years

4. The three main federal government agencies that regulate asbestos are

   A. Food and Drug Administration, Department of Transportation, Environmental Protection Agency.

   B. Department of Transportation, Environmental Protection Agency, Occupational Safety and Health Administration.

   C. Department of Health and Human Services, Environmental Protection Agency, Occupational Safety and Health Administration.

   D. General Services Administration, Department of Health and Human Services, Occupational Safety and Health Administration.

   E. No federal government agencies regulate asbestos.

5. Which of the following are not the responsibility of the Local Education Agency?

   A. Must conduct periodic surveillance in each building under its authority at least once every 6 months and use an accredited inspector to conduct the reinspections every 3 years.

   B. Must attach a warning label immediately adjacent to any friable and nonfriable asbestos-containing building material (ACBM) and suspected ACBM located in routine maintenance areas, such as boiler rooms, at each school building.

C. Must send all notification, inspection, and periodic surveillance records to the EPA on an annual basis.

D. Ensure that complete and up-to-date records of inspections, reinspections, response activities, periodic surveillances, and operations and maintenance activities are maintained.

E. Must comply with the notification requirements to workers, students, building occupants, parents, and short-term workers.

6. Which activities must be conducted by an accredited inspector?

A. Identify all homogeneous areas of material that are suspected to contain asbestos

B. Gather information on the uses and functions of the spaces within the homogeneous areas

C. Collect samples of material suspected to be ACBM and send them to the lab for analysis

D. Perform a physical assessment of the material and document the results in an inspection report

E. All of the above activities

7. Some of the most common uses of asbestos-containing building materials found include

A. Fireproofing on structural members

B. Plaster, pipe, and boiler insulation

C. Acoustical or soundproofing material

D. Flooring and ceiling tiles

E. All of the above

8. In addition to imposing other requirements, the Asbestos Hazard Emergency Response Act requires that a local education agency

A. Close buildings in which asbestos is found

B. Perform inspections to identify asbestos-containing building materials in its buildings

C. Notify the Environmental Protection Agency on the locations of asbestos-containing building materials in the schools of the district

D. Remove all asbestos-containing building materials from its buildings

E. B and D

9. A management plan must contain appropriate response actions. Which of the following is not an appropriate response action?

A. Replace damaged asbestos-containing building materials with new undamaged asbestos-containing building materials

    B. Repair damaged asbestos-containing building materials to an undamaged or intact condition

    C. Encapsulate asbestos-containing building materials with a material that surrounds or embeds asbestos fibers

    D. Enclose asbestos-containing building materials in an airtight, impenetrable permanent barrier

    E. None; all of these are appropriate response actions

10. At least once every month the local education agency must conduct a visual inspection of all areas identified in the management plan as asbestos-containing building materials or assumed to contain asbestos-containing building materials to determine whether the condition of the ACBM or assumed ACBM has changed. This is called a

    A. 12, periodic surveillance

    B. 12, inspection

    C. 6, periodic surveillance

    D. 6, inspection

    E. 24, reinspection

11. Final air clearance of a functional space after a response action to remove, encapsulate, or enclose ACBM involves the following:

    A. Visual inspection

    B. Collection of air samples

    C. Analysis of samples by PLM

    D. Analysis of samples by TEM, unless the project involves less than 160 square feet or 260 linear feet, in which PCM may be used

    E. A, B, and D

12. How can the local education agency best minimize accidental disturbances of ACBM during maintenance and renovation activities?

    A. Establish a permit system that calls for all work orders and requests to be processed through the AHERA designated person

    B. Require the AHERA designated person to maintain AHERA inspector and management planner accreditations

    C. Require the principals of all schools to attend asbestos awareness training

    D. Require all periodic surveillance inspections to be conducted by accredited inspectors

    E. Assure that all AHERA management plans are updated on an annual basis

13. A designated person must
    A. Receive training that provides basic knowledge of a number of asbestos-related subjects, as listed in the EPA's asbestos regulations
    B. Complete an EPA- or state-approved inspector course and become accredited
    C. Have a college degree
    D. Pass an EPA test on Designated Person roles and responsibilities
    E. Complete no training

14. An asbestos management program is subject to which EPA statutes and regulations?
    A. Asbestos Hazard Emergency Response Act
    B. Asbestos Hazard Emergency Response Act, National Emissions Standards for Hazardous Air Pollutants
    C. Asbestos Hazard Emergency Response Act, National Emissions Standards for Hazardous Air Pollutants, EPA Worker Protection Rule
    D. Asbestos Hazard Emergency Response Act, National Emissions Standards for Hazardous Air Pollutants, EPA Worker Protection Rule, and Asbestos School Hazard Abatement Reauthorization Act
    E. None of the above

15. Local education agencies must conduct the following notifications:
    A. Annually to parents, teachers, and employee organizations on the availability of the asbestos management plan
    B. Annually to workers, building occupants, and their guardians on recent or planned asbestos activities (e.g., inspections, response action, etc).
    C. To short-term workers (e.g., telephone repair workers, utility workers, or exterminators) who may come into contact with asbestos on the locations of asbestos-containing building materials (or assumed ACBMs)
    D. Annually to EPA or state agencies on updates to the management plan
    E. A, B, and C

16. The management plan must be
    A. Kept in the local education agency's administrative office
    B. Kept in the administrative office of each school building
    C. Available to persons for inspection without cost or restriction
    D. Complete and up to date
    E. all of the above

Answers

(1) A (2) D (3) A (4) B (5) C (6) E (7) E (8) B (9) A (10) C (11) E (12) A (13) A (14) D (15) E (16) E

Glossary

Active waste disposal site means any disposal site other than an inactive site.

Adequately wet means sufficiently mix or penetrate with liquid to prevent the release of particulates. If visible emissions are observed coming from asbestos-containing material, then that material has not been adequately wetted. However, the absence of visible emissions is not sufficient evidence of being adequately wet.

Asbestos means the asbestiform varieties of serpentinite (chrysotile), riebeckite (crocidolite), cummingtonite-grunerite, anthophyllite, and actinolite-tremolite.

Asbestos-containing waste materials means mill tailings or any waste that contains commercial asbestos and is generated by a source subject to the provisions of this subpart. This term includes filters from control devices, friable asbestos waste material, and bags or other similar packaging contaminated with commercial asbestos. As applied to demolition and renovation operations, this term also includes regulated asbestos-containing material waste and materials contaminated with asbestos including disposable equipment and clothing.

Asbestos mill means any facility engaged in converting, or in any intermediate step in converting, asbestos ore into commercial asbestos. Outside storage of asbestos material is not considered a part of the asbestos mill.

Asbestos tailings means any solid waste that contains asbestos and is a product of asbestos mining or milling operations.

Asbestos waste from control devices means any waste material that contains asbestos and is collected by a pollution control device.

Category I nonfriable asbestos-containing material (ACM) means asbestos-containing packings, gaskets, resilient floor covering, and asphalt roofing products containing more than 1 percent asbestos as determined using the method specified in appendix E, subpart E, 40 CFR part 763, section 1, Polarized Light Microscopy.

Note: Excerpted from *www.slocleanair.org/business/pdf/40cfr61m.pdf*

Category II nonfriable ACM means any material, excluding Category I nonfriable ACM, containing more than 1 percent asbestos as determined using the methods specified in appendix E, subpart E, 40 CFR part 763, section 1, Polarized Light Microscopy, that, when dry, cannot be crumbled, pulverized, or reduced to powder by hand pressure.

Commercial asbestos means any material containing asbestos that is extracted from ore and has value because of its asbestos content.

Cutting means to penetrate with a sharp-edged instrument and includes sawing but does not include shearing, slicing, or punching.

Demolition means the wrecking or taking out of any load-supporting structural member of a facility together with any related handling operations or the intentional burning of any facility.

Emergency renovation operation means a renovation operation that was not planned but results from a sudden, unexpected event that, if not immediately attended to, presents a safety or public health hazard, is necessary to protect equipment from damage, or is necessary to avoid imposing an unreasonable financial burden. This term includes operations necessitated by nonroutine failures of equipment.

Fabricating means any processing (e.g., cutting, sawing, drilling) of a manufactured product that contains commercial asbestos, with the exception of processing at temporary sites (field fabricating) for the construction or restoration of facilities. In the case of friction products, fabricating includes bonding, debonding, grinding, sawing, drilling, or other similar operations performed as part of fabricating.

Facility means any institutional, commercial, public, industrial, or residential structure, installation, or building (including any structure, installation, or building containing condominiums or individual dwelling units operated as a residential cooperative, but excluding residential buildings having four or fewer dwelling units); any ship; and any active or inactive waste disposal site. For purposes of this definition, any building, structure, or installation that contains a loft used as a dwelling is not considered a residential structure, installation, or building. Any structure, installation, or building that was previously subject to this subpart is not excluded, regardless of its current use or function.

Facility component means any part of a facility including equipment.

Friable asbestos material means any material containing more than 1 percent asbestos as determined using the method specified in appendix E, subpart E, 40 CFR part 763, section 1, Polarized Light Microscopy, that, when dry, can be crumbled, pulverized, or reduced to powder by hand pressure. If the asbestos content is less than 10 percent as determined by a method other than point counting by polarized light microscopy (PLM), verify the asbestos content by point counting using PLM.

Fugitive source means any source of emissions not controlled by an air pollution control device.

Glove bag means a sealed compartment with attached inner gloves used for the handling of asbestos-containing materials. Properly installed and used, glove bags provide a small work area enclosure typically used for small-scale asbestos stripping operations. Information on glove-bag installation, equipment and supplies, and work practices is contained in the Occupational Safety and Health Administration's (OSHA's) final rule on occupational exposure to asbestos (appendix G to 29 CFR 1926.58).

Grinding means to reduce to powder or small fragments and includes mechanical chipping or drilling.

In poor condition means the binding of the material is losing its integrity as indicated by peeling, cracking, or crumbling.

Inactive waste disposal site means any disposal site or portion of it where additional asbestos-containing waste material has not been deposited within the past year.

Installation means any building or structure or any group of buildings or structures at a single demolition or renovation site that are under the control of the same owner or operator (or owner or operator under common control).

Leak-tight means that solids or liquids cannot escape or spill out. It also means dust-tight.

Malfunction means any sudden and unavoidable failure of air pollution control equipment or process equipment or of a process to operate in a normal or usual manner so that emissions of asbestos are increased. Failures of equipment shall not be considered malfunctions if they are caused in any way by poor maintenance, careless operation, or any other preventable upset conditions, equipment breakdown, or process failure.

Manufacturing means the combining of commercial asbestos—or, in the case of woven friction products, the combining of textiles containing commercial asbestos—with any other material(s), including commercial asbestos, and the processing of this combination into a product. Chlorine production is considered a part of manufacturing.

Natural barrier means a natural object that effectively precludes or deters access. Natural barriers include physical obstacles such as cliffs, lakes or other large bodies of water, deep and wide ravines, and mountains. Remoteness by itself is not a natural barrier.

Nonfriable asbestos-containing material means any material containing more than 1 percent asbestos as determined using the method specified in appendix E, subpart E, 40 CFR part 763, section 1, Polarized Light Microscopy, that, when dry, cannot be crumbled, pulverized, or reduced to powder by hand pressure.

Nonscheduled renovation operation means a renovation operation necessitated by the routine failure of equipment, which is expected to occur within a given period based on past operating experience, but for which an exact date cannot be predicted.

Outside air means the air outside buildings and structures, including, but not limited to, the air under a bridge or in an open-air ferry dock.

Owner or operator of a demolition or renovation activity means any person who owns, leases, operates, controls, or supervises the facility being demolished or renovated or any person who owns, leases, operates, controls, or supervises the demolition or renovation operation, or both.

Particulate asbestos material means finely divided particles of asbestos or material containing asbestos.

Planned renovation operations means a renovation operation, or a number of such operations, in which some RACM will be removed or stripped within a given period of time and that can be predicted. Individual nonscheduled operations are included if a number of such operations can be predicted to occur during a given period of time based on operating experience.

Regulated asbestos-containing material (RACM) means (a) Friable asbestos material; (b) Category I nonfriable ACM that has become friable; (c) Category I nonfriable ACM that will be or has been subjected to sanding, grinding, cutting, or abrading; or (d) Category II nonfriable ACM that has a high probability of becoming or has become crumbled, pulverized, or reduced to powder by the forces expected to act on the material in the course of demolition or renovation operations regulated by this subpart.

Remove means to take out RACM or facility components that contain or are covered with RACM from any facility.

Renovation means altering a facility or one or more facility components in any way, including the stripping or removal of RACM from a facility component. Operations in which load-supporting structural members are wrecked or taken out are demolitions.

Resilient floor covering means asbestos-containing floor tile, including asphalt and vinyl floor tile, and sheet vinyl floor covering containing more than 1 percent asbestos as determined using polarized light microscopy according to the method specified in appendix E, subpart E, 40 CFR part 763, section 1, Polarized Light Microscopy.

Roadways means surfaces on which vehicles travel. This term includes public and private highways, roads, streets, parking areas, and driveways.

Strip means to take off RACM from any part of a facility or facility components.

Structural member means any load-supporting member of a facility, such as beams and load-supporting walls; or any nonload-supporting member, such as ceilings and nonload-supporting walls.

Visible emissions means any emissions that are visually detectable without the aid of instruments coming from RACM or asbestos-containing waste material, or from any asbestos milling, manufacturing, or fabricating operation. This does not include condensed, uncombined water vapor.

Waste generator means any owner or operator of a source covered by this subpart whose act or process produces asbestos-containing waste material.

Waste shipment record means the shipping document, required to be originated and signed by the waste generator, used to track and substantiate the disposition of asbestos-containing waste material.

Working day means Monday through Friday and includes holidays that fall on any of the days Monday through Friday.

Index

Note: Page numbers followed by *b* indicate boxes, *f* indicate figures, and *t* indicate tables.

Printed in the United States
By Bookmasters